New Security Challenges

Series editor
George Christou
Department of Politics and International Relations
University of Warwick
Coventry, UK

The last decade has demonstrated that threats to security vary greatly in their causes and manifestations and that they invite interest and demand responses from the social sciences, civil society, and a very broad policy community. In the past, the avoidance of war was the primary objective, but with the end of the Cold War the retention of military defence as the centrepiece of international security agenda became untenable. There has been, therefore, a significant shift in emphasis away from traditional approaches to security to a new agenda that talks of the softer side of security, in terms of human security, economic security, and environmental security. The topical *New Security Challenges* series reflects this pressing political and research agenda.

More information about this series at
http://www.springer.com/series/14732

Håkan Gunneriusson

Bordieuan Field Theory as an Instrument for Military Operational Analysis

palgrave
macmillan

Håkan Gunneriusson
Department of Military Studies
Swedish Defence University
Stockholm, Sweden

New Security Challenges
ISBN 978-3-319-65351-8   ISBN 978-3-319-65352-5   (eBook)
DOI 10.1007/978-3-319-65352-5

Library of Congress Control Number: 2017951520

Cover illustration: Pattern adapted from an Indian cotton print produced in the 19th century

Printed on acid-free paper

This Palgrave Macmillan imprint is published by Springer Nature
The registered company is Springer International Publishing AG
The registered company address is: Gewerbestrasse 11, 6330 Cham, Switzerland

To my children, Saga and Axel and my loving wife Christina

PREFACE

Somewhere, there is an armed conflict. Who are "we", who are "they" and what is actually happening in the operational area? If these questions can be answered, one has taken great steps towards being able to predict the unfolding of the events ahead in a conflict. The following book is presenting tools to help address these questions for primarily military organisations, but also for any actor collective or individual dealing with international conflicts. The approach is sociological and the examples are thought to serve as inspiration for others to work foremost with the perspective but perhaps also with the methods. The purpose of this book is to give researchers in the field of war studies and neighbouring disciplines inspiration and guidelines on how to apply field theory on subjects dealing with conflicts and problems associated with conflicts.

Leadership in conflicts are very much based both on experience and predictions, predictions based on these aforementioned experiences. These experiences are important in that they have shaped the range of perspectives which the actor (collective or individual) has, not all experiences are equally important when it comes to making an imprint. This is natural and often leads the actors correctly, if not they sometimes have room to adapt, sometimes not. This is the logic of practice through the ages of military leadership. On the other hand, one of the oldest institutions in Western Civilisation is the institution of universities. Since the Enlightenment the universities have been in the service of bringing knowledge into all areas of activities within states and society. Still, socio-logical perspectives on warfare have often been limited to studying the conflict from an outside perspective without primarily being interested in delivering actual tools for the operators engaged in the conflict.

Stockholm, Sweden   Håkan Gunneriusson

CONTENTS

ABOUT THE AUTHOR

Håkan Gunneriusson (Ph.D) is the Head of research/Deputy Head of Land Operations Section at the Department of Military Studies at Swedish Defense University (SEDU) focusing on hybrid warfare, military ground tactics, as well as sociological and historical perspectives on military tactics and culture. He has been teaching and researching at multiple universities and teaches not at all levels of the officer's programs at SEDU.

ABBREVIATIONS

BiH	Bosnia and Herzegovina
COIN	Counter-Insurgency
DIME	Diplomacy, Information, Military and Economy
EBO	Effect Based Operations
HDZ	The Croatian Democratic Union (*Hrvatska demokratska zajednica*)
HUMINT	Human Intelligence
IFOR	Implementation Force
INFOPS	Information operations
JFCOM	United States Joint Forces Command
JNA	Yugoslavia's National Army
MCA	Multiple correspondence analysis
NATO	The North Atlantic Treaty Organization
NGO	Non-Governmental Organisations
OSCE	(Organisation for Security and Co-operation Europe)
PIRA	Provisional Irish Republican Army
PSYOP	Psychological Operations
SDA	Party of Democratic Action *(Stranka demokratske akcije)*
SDS	Serbian Democratic Party (*Srpska Demokratska Stranka*)
SFRY	Social Federal Republic of Yugoslavia
SPO	The Serbian Renewal Movement (*Srpski Pokret Obnove*)
SPS	Socialist Party of Serbia (*Socijalistička partija Srbije*)
SRP	*The Party of Serbian Unity (Stranka Srpskog Jedinstva)*
UNPROFOR	UN-Protection Force

LIST OF FIGURES

PART I

Theory and Background Field Theory as an Instrument for Operational Analysis

CHAPTER 1

Terms in Search for a Theoretical Definition

Abstract This chapter deals with why social theory can be relevant in the context of military operations. The chapter does also give some examples of what have been done before in this matter, specifically the concept Effect-Based Operations (EBO). The French sociologist and theorist Pierre Bourdieu is also introduced here.

Keywords Effect-Based Operations (EBO) · Pierre Bourdieu · Algeria

The French sociologist Pierre Bourdieu (1930–2002) first developed his theories about social fields when he made field studies in the North African mountain ranges, among the Kabyle in Algeria. The time period was the 1950s to the early 1960s, when France fought their war in Algeria. Even though most of the Kabyle are Muslims they are not Arabic, and they speak the North African Berber language *kabyle*. He then found that he could only understand them fully if he decoded the symbolic values which they placed on things, customs and behaviour.[1]

---

[1] This text is an extension built on a synopsis paper published in "Fältperspektiv. An approach to achieving one's aims in armed operations" in *Studies in Education and Culture dedicated to Donald Broady*. Ed. Börjesson et al. Uppsala, 2007. For more on both the background to the theory and the theory it-self see: http://www.skeptron.uu.se/broady/sec/ske-15.pdf.

© The Author(s) 2017
H. Gunneriusson, *Bordieuan Field Theory as an Instrument for Military Operational Analysis*, New Security Challenges, DOI 10.1007/978-3-319-65352-5_1

This was a young researcher still strictly empirical, creating the foundation for his theories. Just recently his book *Algerian Sketches* was released post-mortem where he writes about Algeria, the Algerian conflict and his theories. Later on, Bourdieu found that much of the social mechanisms he had found among the Kabyle were present in all manners of other social contexts.

Bourdieu was one of the foremost French thinkers of the late 1900s and had great influence on many humanist and social science disciplines—mainly in Europe. Those who have not read his work consider his thinking to be post-modern, but strictly speaking, he is not of the deconstructive school. He saw himself as post-structuralist, which was partly an acknowledgement of structuralism, but which also showed that it was not as embedded in his thinking with the degree of rigidity normally imposed by structuralism. He has, for example, done extensive research in the fields of art, literature and even studied his peers, the professors in Paris as empirical material. Bourdieu might at times be hard to understand, but his theories are based strictly on empirical material which he then has generalised to theoretical systems. With the passage of time, he introduced new terms, for example the term *field*, which we will return to later on.

For more than a decade, I myself have used and expanded Bourdieu's theories and found that there is a case to state that they are valid for studies and practice regarding Military Operations as well. This text should be seen as an introduction of the theories into the military field, especially regarding the field of intelligence analysis. In military context, the theories can be used to make social patterns visible. If you have good empirical material and a good analysis, then you will be able to see how actors (collective as well as individual) relate to each other. In the end, one can be both able to predict what will happen (with a certain degree of certainty) and able to make the other part (enemy or not) behave in a certain way, without them knowing that you are manipulating them.

I would like to underscore that this text is primarily focused on the theory and not the empirical aspect. It only serves as a way to exemplify the theory and show its usefulness in operational analysis. I recommend the bibliography at the end for those interested in the empirical events briefly explained in this text. It should be stressed that field theory is more of an object than a tool in the text. The text is primarily a tool for training, inspire primarily military officers how to use field analysis. The overall aim of this text is to present an alternative approach to knowledge support. It will, in essence, explain field theory in context and show the theory's potential for helping to understand certain types of military conflicts. This will be achieved by using the scientific theoretical system to create a picture of how an area of operations can be understood from a theoretical perspective. This will involve making a science of military practice, by transforming scientific theory into a military modus operandi in the operations area.

In counterterrorisms theory with roots from the 1970s, you have an enemy centric approach, whereas the so-called classic counterinsurgency has a more population centric approach.[2] Both have an actor first perspective and do not focus on the structure. A post-structuralist approach will be used here, which is a mix of actor and structure perspective. Methods will also be examined for the creation of a social field. However, a field theory analysis based on an actual conflict will not be conducted. A completely different scale of effort would be required for this; the conflict itself would have to be in progress and considerable operational resources would be necessary. Instead, an example—the former Yugoslavia—will be used to show the feasibility of using field theory to understand a low intensity conflict. The reason why this discussion will concentrate mainly on the strategic level (politico-strategic and military strategic) is a result of the relatively good accessibility of source materials. However, it would be reasonable to consider using field theory as an operational and tactical tool, but the source material required to produce illustrative examples would amount to something approaching a collective biography of the many actors involved.

In order to generate a theoretical discussion around operations of so wide-ranging nature as those in play when a country or a region is threatened or finds itself at war, requires a balance between generalisation and operationalisation. Theory must become a practical tool to assist those responsible for operational decisions. At the same time, it should remain at a level general enough to allow theory to link to various empiric scenarios without losing relevance. In this text, field theory will be applied to an area where it has not been used before, with empirical material being analysed using a field perspective. After a certain amount of refining, the theory becomes an even sharper tool for the analysis of armed conflict, in particular, peace promotion operations and operations

---

[2] Kilcullen (2009, p. xv).

where information operations are prioritised. This approach will also require considerable information resources from the units involved. However, the aim is not to provide the definitive answer as to how these types of missions will be conducted. The aim is to offer a tool that will more easily enable the successful achievement of tasks for all types of organisations working in the area of operations. The minimum requirement here is for the theory to serve in a teaching role and as an eye-opener for those who have previously not thought in wider conceptual terms. Ideally the model should move from its teaching role to become a practical tool for analysis. As an extension doctrine could be developed based on field theory, in particular for low intensity conflicts such as peace promotion operations or conflicts where manipulation of the social field might provide a fruitful avenue of approach.[3]

Before proceeding further one must acknowledge that a theory can never fully reflect reality, only certain aspects of it. The principal purpose of a theory is to identify the essential elements considered peculiar to the area under study; this is important because reality can often appear far too complex and confusing. What is essential or not is both dependent on what is characteristic for the researched object, but also what at the same time is important for the mission itself. The field of operations described below is not an empirical reality but a heuristic tool used here to provide a foundation for decision makers. The theoretical model avoids going into detail, it merely describes an approach to an empirical situation. In this respect, the theory is well-suited to its purpose in that it does not lay down any detailed guidelines; it is the empirical reality that directs the practical design of the model. For example, the definition of what is important for the actors in this study is based entirely on the assessment of the actors themselves, the assessment is *not* shaped by theory nor is the nature of the theoretical categories determined on beforehand—more on this later.

When it comes to the insurgency type of warfare, we in the *West* have had the winning of hearts and minds approach in some shape or form at hand since the Vietnam war. The concept of Strategic communication

---

[3] "Low intensity conflict" is a vague term, but it may be defined as conflict that does not fulfil the criteria for war. The latter is itself a term with several definitions, but that low intensity warfare is different from high intensity warfare (i.e. war) is often considered to be true. It is the criteria that changes, not the relationship between the terms. *Militärstrategisk doktrin* (2002), p. 103.

has been important and still is. Prominent authors like David Kilcullen and John Nagl have been proponents of this approach, even if they of course also realise that it is a problematic endeavour.[4] A problem with strategic communication is that it is more of a strategic monologue. The concept being put forwards in strategic communication can be as basic in practice as building roads but should deliver the message that there is righteousness in our concept of why the conflict is fought. If we take the opposition's ideas into account we might need to find a middle ground and that is not desirable, as it delegitimises the very idea of the conflict at home. Strategic communication is basically a monologue because the communicator wants it to be one and by that, it is often not very successful as it does not appreciate the Clausewitzian concept of the conflict being a duel situation. The handling of the Iraq situation after the toppling of Saddam is a clear example of a strategic monologue with disastrous effects. There was no willingness to read the situation and act accordingly, it was just a dictate and efforts to push that dictate out to the population.[5] Others have underscored that the USA must improve when dealing with HUMINT on the tactical level.[6] It is perhaps stating the obvious as improvement is always wanted, the real question is *how*.

It might also prove useful to touch on some of what has been written about the now less used military concept EBO (Effects-Based Operations).[7] This paper will not deal with EBO, but rather serve as an alternative to it. The theoretical foundation for EBO was so hollow yet so widely discussed that it is better to leave that debate open and concentrate on developing our theoretical thinking on military operations instead.[8] But the EBO debate can serve as an example that there was

---

[4] Killcullen (2006, 2009, 2010), Nagl (2005).

[5] Forbidding even low-level Baath party members' state employment, disbanding the whole Iraqi army without compensation and not recognising Islam as a political force was an almost perfect recipe for disaster.

[6] Cordesman (2004, pp. 40, 44, 51).

[7] Mattis (2008).

[8] When the term EBO was in fashion there was a wide range of interpretations. Ho (2005, pp. 64).

and actually still is need for theory regarding military operations. A number of the articles written attempted to capture the essence of the new term, which at the time was afforded a certain currency, with the aim of loading the term with old rehashed material so that more resources could be obtained for projects initiated earlier.[9] An approach that resonates with the content of this paper is offered by the researchers Michael Callan and Michael Ryan: "Effects-Based Operations are the application of military and non-military capabilities to realise specific and desired strategic and operational outcomes in peace, tension, conflict and post-conflict situations". The later application of the term maintained a relatively general level, which was appropriate since the approach required a high degree of generalisation. The definition is also good because it can be applied to the border area between the use of force and more peaceful means, as described in the above text. The fact is that the definition works very well from a field theory perspective underlines the relevance of the field theory approach. The focus below is on those aspects of the JFCOM (United States Joint Forces Command) interpretation that stress the winning of peace.[10] The aspects of both interpretations that focus on weapons effects are not relevant for the purposes of this paper.[11] The theory presented in this text offers a practical approach that can be applied in situations where conventional warfare and its concepts are not appropriate. The issue is scientifically relevant, much research, for example, having been conducted into examining terms such as the now dearly departed EBO or the current strategic communication approach. This particular term is one for which researchers and other writers have created a number of definitions, the problem having been very much one of mastering the terminology. This resulted in a series of straggling

---

[9] See, for example Dawen (2005, p. 81) and Herndon et al. (2005). Stockholm.

[10] (Wikström 2005, p. 12).

[11] The theoretical writings on this topic are somewhat undeveloped and hackneyed. See, for example David A. Deptula's paper Effects-Based Operations. Without going into detail it focuses on what action one can take against an enemy that is incapable of defending himself. The study also gives an insight into the competition that takes place between the armed services in the USA.

definitions, simply because the term EBO was formulated without any normative content. One often has an idea of what one would like to achieve on a particular operation, but there has been no theoretical framework to enable a link to the problem at issue. One approach is to create specific theoretical terms to facilitate analysis of military operations. However, it might also be of value to do what is often done in academic circles, to apply a fresh, untested line of thinking to the material under study. This is the approach that will be adopted in this text, albeit only purely tentatively. This shows that it is far from impossible for terms to be given value without even the definitions of the term itself being defined, it is like a casino where the stakes are being thought over and the winner defines the content of what was at stake. A more reasonable approach is, of course, to define what is being discussed before proceeding with a power struggle over the same subject of discussion.

A theoretical perspective will now be presented that gives an alternative perspective to military planning, or it will at least provide an understanding of how an operations area can be analysed using theory. It is worth stressing that this does not necessarily mean presenting a theory that will lead to new practices on the field. Rather it so happens that much of what is advocated by the theory actually already occurs on the field. The problem is that the practices being examined here have hitherto lacked any form of explanatory foundation, other than that proven experience has shown that they work well. If the practices are given a theoretical explanation this may illuminate how current practice can be further developed. Therein lies the benefit of a theory that can be applied to the practices under discussion here.

I emphasise this point here, but believe it to be so important that you will find it repeated throughout the text: theories are used mainly to *generate issues of interest* that will be played out empirically. *Theories are not primarily used to provide answers to questions.* If the latter were the case empirical research would not be necessary, theory alone would suffice to explain reality. This is an unempirical process that should be avoided. Only in exceptional cases where there is a lack of empirical foundation can one generalise using an empirical approach, and assume that a situation will play out in a certain way. Nonetheless, theory is important in empirical research, as it helps us structure the reality which empirical data consists of.

Military intervention in an area where armed conflict is taking place is, to say the least, a risky undertaking. There is a mass of information that decision makers can and must gather and analyse in order to find answers to important questions. In addition, there is a wealth of information that cannot be acquired beforehand. Once a military force has intervened in a crisis area the conditions will change and new structural patterns may emerge. However, it is of great help if one has a theoretical model in advance that can be applied to the local arena, especially if the same picture is shared by everyone in your organisation. This essay introduces such an approach and aims to show the value of field theory. Military operations will always be conducted in line with a specific operational practice that will often have been well tested. Why does this practice take this particular form? A simple answer is because it has been shown to work. But why does it work? What are the underlying structures on which that practice is based?

The method in this text has a disadvantage in a quickly evolving scenario if the theoretical approach is to be used in a quickly unfolding ongoing conflict, by, for example, military units. The method here in is the one which one rather uses after the conflict, in order to try to reconstruct the events. Or it can be used in a prolonged conflict where there is time for analyses as events unfold. One can classify the method as a qualitative analysis based on primary and secondary sources.

The method chosen, however, is to show through the use of a number of examples how sociological theory can be applied in general to military planning. In particular, the so-called *field theory* is put forwards as the example, a theory hatched by Pierre Bourdieu. This theory, after examination, will then be applied to the scenario of the dissolution of Yugoslavia. The methods used in a real context depend primarily on what kind of source materials there are. The theory is not dependent on any specific type of method, so one can use whatever one thinks works best, qualitative as well as quantitative material. This and the following text present a theory that will be transformed into practice; it is both a generalisation of reality and a road map for practice in the same reality. The preferred method for a contemporary ongoing conflict would in this case be quantitative correspondence analysis often called *MCA*, and it is a statistical method. For those interested in this method one can look at the course material for a wide range of courses teaching this method.[12]

---

[12] Recommended software for performing the method can be found here: http://www.spadsoft.com/ [Visited 170509].

The method is best explained in the book *Multiple Correspondence Analysis* by Birgitte le Roux and Henry Rouanet.[13] This method makes use of a large host of different data, of which you do not know the real importance at the start of the case—it is a *tabula rasa*, to get the value of the data mapped during the research. You can, for example, use 50 different questions just about religion, they can be keywords, or physical objects like buildings, or religious practices. One can suspect that some of them have importance and some do not, but confronted with the actors on the field one might get some surprising results which let you understand the field as a whole together with all the other parameters (perhaps thousands) you confront the reality with. You get a pattern, a system of beliefs which constitute the field with this method.

BIBLIOGRAPHY

Cordesman, Anthony H. (2004). *The Military Balance in the Middle East.* 25, 525, Westport: Praeger.

Dawen, C. (2005). Effect-based operations: Obstacles and opportunities. In J. Elg (Ed.), *Effektbaserade operationer.* Stockholm: Swedish National Defence University.

Försvarsmakten, (2002). *Militärstrategisk doktrin.* Stockholm: Försvarsmakten.

Herndon, R. B., et al. (2005). Effects-based operations in Afghanistan: The CJTF-180 method of orchestrating effects to achieve objectives. In J. Elg (Ed.), *Effektbaserade operationer.* Stockholm: Swedish National Defence University.

Ho, J. (2005). The advent of a new way of war: Theory and practice of effect based operations. In J. Elg (Ed.), *Effektbaserade operationer.* Stockholm: Swedish National Defence University.

Kilcullen, D. (2006). Twenty-eight articles: Fundamentals of company-level counterinsurgency. *Military Review, 83*(3): 103–108.

Kilcullen, D. (2009). *The accidental guerrilla. Fighting small wars in the midst of a big one.* Oxford: Oxford University Press.

Kilcullen, D. (2010). *Counterinsurgency.* Oxford: Oxford University Press.

Mattis, J. (2008). Commander's guidance for effects-based operations. *Joint Forces Quarterly, 51*, 4. (Washington, DC).

Nagl John A. (2005). *Learning to Eat Soup with a Knife: Counterinsurgency Lessons from Malaya and Vietnam.* The University of Chicago Press.

---

[13] Le Roux and Rouanet (2010).

le Roux, B., & Rouanet, H. (2010). *Multiple correspondence analysis.* London: Sage.
Wikström, N. (2005). Introduktion: Effects-based operation, en högre form av krigskonst. In J. Elg (Ed.), *Effektbaserade operationer.* Stockholm: Swedish National Defence University.

CHAPTER 2

Field Theory

Abstract In this chapter, the theoretical framework of field theory is presented. The freedom of framing the field depending on the task is discussed along with terms as capital, autonomous vs. heteronomous and illusio. An example of a social field of the discipline of history of war is also presented.

Keywords Field theory · Capital · Illusio

Field theory as mentioned above can be used to describe a practice by the military, a practice that very often can be quite innovative. Field theory has originally nothing to do with *the field* in the military sense. However, what field theory is able to provide is a theoretical and comprehensive explanation of the logic of practice in the area of operations, which has hitherto been lacking. Consequently, old practice can be placed in a new context and given relevance in situations that have not been previously addressed; new systems become available suggesting a range of different modi operandi.

One can compare the relation between practice and theory to the ever-current scientific theoretical discussions on the relationship between technology and science, which seem to follow one after the other. There is no given answer, but the question becomes of interest when it generates scientific problems. It is often the case that technology precedes science, the purpose of science after all being to explain the overall context

H. Gunneriusson, *Bordieuan Field Theory as an Instrument for Military Operational Analysis*, New Security Challenges, DOI 10.1007/978-3-319-65352-5_2

of the results of practical technology. Field theory can fulfil the same function in the military arena. Adopting this approach with the explanatory perspective of theory guiding the way, the theory enables field operators to take things forward in a normative fashion.[1]

Field theory has proved to be a successful tool in a multitude of studies over decades, focusing on cultural factors. Questions such as which factors are or are not important for a particular group of actors are typically those that field theory can be used to answer. The answers are of great importance because for one's own operations, it will be essential to understand the operational theatre in general and the opposing parties in particular. Armed conflict is a battle of wills. To hold firmly to one's course in the face of opposition may of course lead to subjugating your enemy, but to win the peace requires more than just subjugation. You need to influence the opposition's perception of what is important and not.

Field theory is probably most appropriate for prolonged low-intensity conflicts. This, however, is an empirical question, but it is probably more productive to use field theory and methods other than military ones to try to understand and influence the opposing party when the conflict is still in the peace promotion phase. There is apparently no lack of definitions associated with low-intensity conflict, quite the reverse.[2] Low-intensity conflict presents an arena in which field theory could enable a smoother achievement of designated aims. That said, I now intend to avoid discussion on the definition of low-intensity and other conflicts. The not unusual prolonged length of low-intensity conflict makes field theory a viable option for analysis as it takes time to gather and process data in a heuristic process.

The issue of definitions is principally of academic interest. For those interested in achieving a successful outcome in the operational theatre, it is better to start from the other end: conduct an empirical study of the conflict, determine what needs to be achieved and thereafter decide if intervention is necessary and if so, what kind, e.g. conventional warfare or more peace-oriented operations. In other words, one has to decide whether the armed forces in question will go in with the most advanced

---

[1] For a discussion on normative and descriptive theory in military theory, see, for example, Ångström (2003, p. 154).

[2] Ångström (2003).

weapon systems at their disposal, e.g. tanks, electronic warfare units, cyberwar, aircraft carriers, fighter aircraft, air defence systems, heavy artillery, or if just boots on the ground will suffice. Once that decision has been made the definition will follow in line with the course of action chosen.[3] Defining the level of conflict has no intrinsic value; it is the operational effect that is the crucial factor.

Field theory does not provide a means of bending the enemy to one's will, rather of getting the opposing party to voluntarily change strategy to one better aligned to one's own aim for the operation. The deterrent effect achieved by a show of armed force is certainly one component of the theory, but it is not the most important factor. In his book, *Arms and Influence,* Thomas Schelling developed the concept of the deterrent effect associated with an armed force capability.[4] Schelling is no longer alone in this field, with a complete genre of literature on the topic of diplomacy backed by the threat of armed force now in existence.[5] The Nobel laureate Schelling is, however, a prominent portal figure in this field of research.

Possessing the potential for armed force has a deterrent effect and is certainly an important factor, but securing the monopoly of armed force is only a minimum requirement not an optimum criterion.[6] If field theory is to be linked to military thinking, then Sun Tzu is the one who best captures the essence of the theory.[7] He believed that a strategy that delivers victory without the need for battle is the supreme strategy that will win all and many battles. This is also the heart of field theory, with the focus not on defeating the enemy. Neither is field theory a question of forcing the enemy to choose his next best strategy, or even a worse strategy. Field theory is about changing the conditions for all actors in the operational area to ensure that the strategies that do not involve the use of force are those that will most easily lead to the achievement of the goals established by the actors.

---

[3] One might consider that a misjudgement was made by the Armed Forces, and if so then a redefinition is obviously required.

[4] Schelling (1996). The reasoning in Chap. 5 is particularly inspiring in this context.

[5] See, for example, Alexander L. George's excellent paper The Limits of Coercive Diplomacy, which places considerable emphasis on measures to encourage change and not just on threats to use armed force. The book also presents several examples of research into this genre.

[6] Weber (1989, p. 64).

[7] Sun Tzu (1997, p. 25).

The differences between Schelling's line of argument and the above are many, but they can be summarised in simple terms. Schelling wants to coerce the opposition to choose an alternative other than his preferred alternative. Field theory in this application centres on changing *the structure of the social arena* so that the opposition's preferred strategy is one we find acceptable and can therefore allow them to adopt. However, the most important thing is that the opposing party is encouraged to pursue the new strategy, to abandon the previous practice by changing the logic of the field. The focus for operations is thus to change the political conditions, using both the whip and the carrot, and is primarily not concerned with the duel between the actor in question and one's own camp. The dangers of focusing on one of the actors and not the whole arena are at least twofold:

(1) There will always be a large number of actors, and by focusing on their environment (both the political and the physical), one will be able to influence them all. If one chooses to concentrate one's focus solely on one or a handful of actors, the others will get away, and one will not therefore be able to take a comprehensive overview of the situation.

(2) The other problem with focusing on individual actors is that this approach will result in the ultimate goal being viewed through a different lens from that which would have been used if one had focused on the arena as a whole. This is a problem which can lead to the wrong decisions being taken, because the essentials and the non-essentials have not been kept separate. There are problems and, of course, opportunities to explore, but the text is mainly tutorial in its character.

Should one therefore, in line with the above argument, completely ignore the opposition? No, it is a question of which priorities to adopt, the arena as a whole taking priority over individual actors. The individual actors must, however, be allowed to change their strategy. If this approach is to succeed, a detailed qualitative knowledge of the actors and their agendas must be accumulated. What are the actors' agendas? To which alternative strategies should they be allowed to change? Can a collective actor be allowed to spread strategy over a more acceptable range of agendas? The main aim, however, is to change the local conditions in the operations area, to create a situation where the local actors, of their own volition, change to a strategy more aligned to one's own aims.

An operations' area may be described as a cultural field, which in its turn is a mutating social space. The theory itself embraces a battery of terms and ideas which are interesting from a meta-perspective. However, for the present, we do not need the entire complex of field theory to create a theoretical platform for understanding warfare. Inspiration has of course been drawn from existing theory when called for. *Culture* in this context does not refer to what in everyday language we would call *highbrow culture*; it has instead a broader meaning. Culture encompasses human activity, specifically in this case human relationships, formal or informal. The term *field* can best be described as a social arena with its own unwritten rules, where the rules set the standards for people's behaviour. These rules are by Bourdieu called *illusio*. The social arena, and perhaps also the geographic arena, comprises only part of the domain covered by an area of operations. The field can also include the current political activity in the area, which is the focus of the following text. The basic values of the active actors (specific individuals, groups or organisations) on the field will determine the future shape of the field. A field may be defined as broadly as the *Balkans' political field*, or as narrowly as the *Mostar political field*, or even more narrowly if the task in hand requires it. The business of deciding what you choose to call a field is governed entirely by the issues and the tasks you are faced with. The term *field* is a theoretical construction based on empirical fact. A field is defined by the gains that one stands to make therein. What are the field's specific values, the stakes involved and the boundaries? The answers to these questions are best provided by those who, during the course of their lives, have lived and worked in that particular field. A field is defined by its characteristics and laws. Each cultural field has its own rules, *illusio*, which in turn defines what will be regarded as *capital*. If one understands the characteristics of a field, then one can form a picture of it, as well as of the agendas of the actors involved and the merits of each position held.[8] Actors are also structured on an individual level, not only on a field level by their illusion. All the experiences of importance (that is subjective of course) an individual has will shape them so

---

[8] Bourdieu (1992, p. 41).

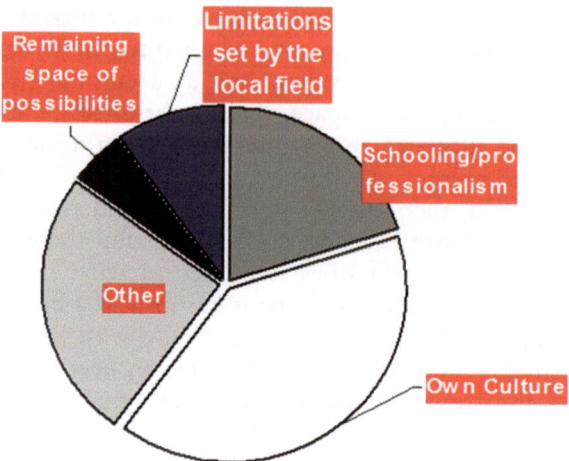

Fig. 2.1 A visualisation of how *habitus* can structure the actor, but also leave freedom of choice within a space of actions. The percentage distribution is purely hypothetical. Not many viable options remain when matching our habitus with what is considered acceptable from the local perspective. It is fully possible to act within the limitations set by the local field; the problem is that such actions do not work very well; 360 degrees represent the theoretical full span of actions a human can perform in a given situation. In the end, a sector is open for the actor to act comfortably within, the width depending on habitus and the situation.

that their future decisions will be decided by their past experiences along with the nature of the situation. Pierre Bourdieu believes that people have the freedom to act within the frame set by their collective experience. He calls that *habitus*. This is a crucial key for an intervention force that intends to influence the very structure of the field in a controlled way. The field will be affected in any case by an intervention, but as a military force, you really want it to happen in a way you can predict (Fig. 2.1).

To be a respected actor in a field, one must understand and accept the field's rules. The unwritten rules of the field are an important part of how the field is defined, which means that the field will take a different form if new rules begin to apply. If new rules are introduced,

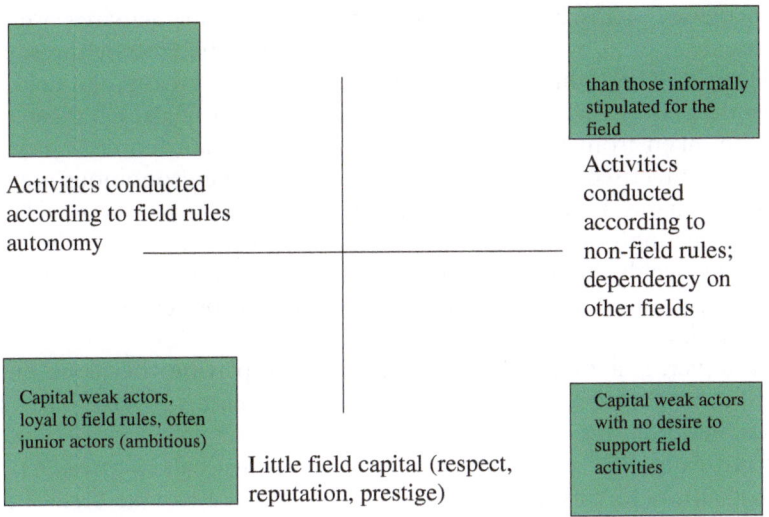

than those informally stipulated for the field

Activities conducted according to field rules autonomy

Activities conducted according to non-field rules; dependency on other fields

Capital weak actors, loyal to field rules, often junior actors (ambitious)

Capital weak actors with no desire to support field activities

Little field capital (respect, reputation, prestige)

Fig. 2.2 A basic sketch of how the four fields of a social field are usually constructed

earlier actors may find their positions threatened. Thus, an international force that is clearly out of touch with the local rules of the field can be perceived as threatening, regardless of the tactics they decide to adopt. There is nothing that says that the rules that apply today will do so tomorrow. This applies most especially to the right to exercise force, which will certainly be restricted for local actors. However, showing a lack of understanding of the field rules will also alienate other groups, who may potentially feel their power base threatened by an unaware newcomer on the field. This poses a constant problem for actors who wish to establish themselves in a social field, for example, in the case of international civilian as well as military intervention (Fig. 2.2).

The many actors in a field also have their own agendas. What holds them all together is that they all believe that the *game* is worth playing. They all adhere to what the field represents and that in turn keeps them as actors on the field. Thus, there are similarities between the actors, but also differences. The actors on the left side lie in the *autonomous* area of the field. In this area, activities are conducted based on the premise that the activities themselves have value. On the right-side activities are conducted motivated by other reasons than the rules dictated by the specific social field. The right side shows the *heteronomous* area of the field, the dependent area, that is: dependent on other social fields. An illustration may be taken from the field of art, where in the autonomous area, art would be practised for art's sake, whereas in the heteronomous area, art would primarily be produced to make, for example, money to take the most occurring example. In many cases, the former art is the one which eventually yields the most money. But the latter category of art can often be mass-produced and really have no ambition to be recognised as art in the first place.

However, it is often the case that someone playing the game on the field has not always made a clear, conscious decision to take part. This is not illogical since the actors see their lives set in a complex reality and not as a game on an abstract field. An individual, as a result of his background and current situation, may be inclined to take certain decisions. These, viewed from the theoretical perspective, are seen as the consequences of the field's structure and the personal dispositions of the actors. The person in question probably does not see it as taking a decision; he just sees it as *the right thing to do*.

An example of a field is the academic historical field, the cultural field for the production of history.[9] The *historical field* exists because a great many of the actors value and are interested in "history". By history, one means its creation in media form, which in some ways could be said to be presenting a testimony to the past. The actors will often have differing opinions on what is considered a faithful and relevant representation of history. As already stated, what they have in common is that they have each decided to remain in the arena. A dislike of someone else's presentation of history is a constituent part of the field's activity, as is to the same degree sympathy with the ideas of others. All of them believe that they are fighting for something important when they directly or indirectly assert their opinions on what they consider to be a

---

[9] The historical field here refers to the field where history is produced, not where it is absorbed—certain rules and logic apply here. A clear example may be the phenomenon whereby literary critics very often do not favour books that are bestsellers.

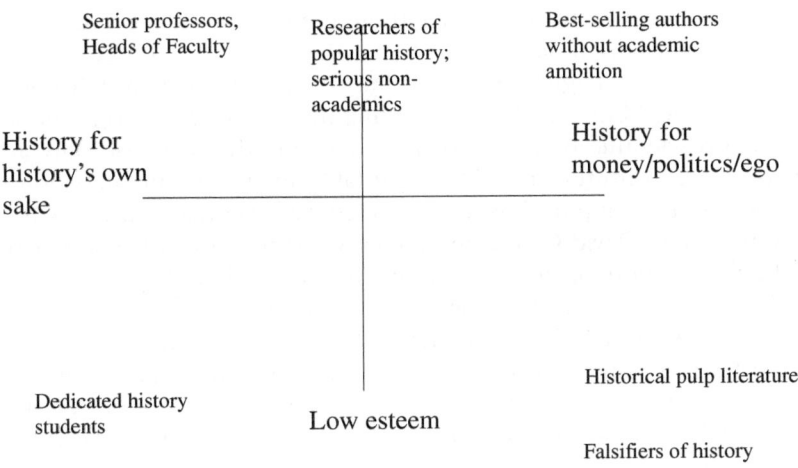

Senior professors,
Heads of Faculty

Researchers of
popular history;
serious non-
academics

Best-selling authors
without academic
ambition

History for
history's own
sake

History for
money/politics/ego

Historical pulp literature

Dedicated history
students

Low esteem

Falsifiers of history

Fig. 2.3   An example of a historical field

faithful rendering of history. For the sake of clarity, a field is presented below showing a selection of what might be included in a historical field (Fig. 2.3).[10]

As the diagram above indicates the field is made up of a disparate collection of actors and activities. Here, it functions more like a *mind map* than an actual positioning based on quantitative data. Holocaust deniers and history professors after all have little in common. But nevertheless, as is shown here, they are all players in the historical field. This shows that the game is one worth playing, although they will each have their own definition of "good" and "bad" history. In this way, they are all part of a whole, the professor probably regarding the falsifier as an out-and-out liar and the history falsifier seeing the professor as someone who has been bought by the establishment.

Even within each of the quadrants, there is considerable breadth. In the top-right quadrant, one would find authors concentrating on military history, such as known names as John Keegan, and Martin van Creveld.

---

[10]For more discussion on the historical field: Gunneriusson (2002a, b). For a more detailed study of the anatomy of a field, Bourdieu (1996a), especially, p. 121, is recommended.

They both are widely read, even though van Creveld devotes himself more to analysis (left side of quadrant) than Keegan, who concentrates more on opinions (right side of quadrant). This leaves them both placed high in the field but with Keegan tending more towards the right because of his non-scientific background. It is not only the range of journalistic ability that constitutes the field. The establishment of reputations on the field is in fact an empirical issue. Respected war historians, such as Charles Edward White, David Glantz and James S. Corum, are not as widely read as the two authors named above, but their exalted academic standing as professors of military history means that their opinions carry considerable weight. They would without hesitation be placed high up on the left side of the field, mainly because of their formal positions as professors. That position serves as a form of acknowledgement, but even a professor can squander his capital. A once-respected war historian who is widely read is David Irving. When over the course of time, it became known that his presentation of history had a political stamp which was in conflict with striving towards objectivity, respect for him declined.[11] From having held a position very similar to Keegan, he is now grouped together with holocaust deniers such as the French literature Professor Robert Faurisson.

The most distinguished professors are placed in the upper-left quadrant of the field, quite irrespective of what these individuals think of each other. Hypothetically speaking, Corum, Glantz and White may disapprove of each other for both personal and professional reasons—just because one's research is sound does not mean that other reliable researchers will agree with the conclusions presented. They have, however, built their positions on the same foundation, by conducting research into military history for its own sake and acquiring such esteemed reputations (symbolic capital) that their word almost amounts to law. Both strive for objectivity, even though there will always exist a certain degree of subjectivity in every position presented. If the most-respected cultural royalty of the upper-left quadrant identifies a phenomenon on the field, then their definition of that phenomenon will be accepted. They have the power, at least partially, to influence field definitions to a disproportionately strong degree merely by voicing their opinions. Conflict between individuals in any given sector can be fierce; they do not need to agree with each other because their positions lie close,

[11] Guttenplan (2005).

but they have achieved the regard in which they are held on the field because they have each accumulated symbolic capital of equal value.

Capital is a word that implies a *resource* that actors have at their disposal. Those who possess this resource have power. The term can be divided into two main categories: symbolic capital and social capital. Social capital may comprise good personal contacts and social networks. For the purposes of this study, the symbolic form of capital is of greater interest. It can be defined as "that which social groups recognise as of value and to which they ascribe value".[12] The term symbolic capital can be seen as a collective expression for prestige, a good reputation, respect and authority. How an individual acquires a good reputation within a particular sphere, profession or in the workplace, is not entirely clear. Many forms of capital are specific to their fields to such an extent that the capital will be afforded little or no acclaim outside the circles that constitute that particular field. Within the subcultures of young people, there are many examples of this. To be able to do a trick with, for example, a skateboard can be the key to the respect of your friends. The same trick will hardly be worth much in the job market or as an aid to gaining better grades at school. The military world is another section of society that has a wide range of capital that can hardly be said to hold much worth in society as a whole. To have or have held a particular appointment is something that is met with a special respect among military personnel but the value of which other people will find hard to understand. The same applies to having served abroad on a particular mission or having studied at a foreign military college of good repute.

The value of symbolic capital is therefore decided by how it is recognised by those who form the immediate environment on the social field. In this respect, a degree of relativity thus permeates the whole field. The majority of the field therefore acknowledges certain types of capital as prestigious and therefore valuable, and the individual actor can extract advantage from his capital if it registers on the scale of values applied by the field. For example, resort to the use of violence is not considered a legitimate option if other political alternatives are available. This will apply if the field is constituted in a similar way to a Western domestic political field. This can, however, change if leading actors on the field advocate other types of capital, for example the capacity for violence. There may also be forms of capital on the

---

[12] Broady (1989, p. 169).

field to which all actors do not have access. This may involve types of prestige and reputation that are not compatible: to act as a representative for different religious groups at one and the same time is seldom possible. Even if all forms of capital are theoretically not available to all actors, there is merit in undermining these inaccessible positions, since power on the field is relative to the other actors participating. For example, it may be difficult to acquire respect as a self-assumed guardian of mosques in Hercegovina if one is at the same time a recognised Croatian militia leader (a position which itself has symbolic value). For this particular militia leader, a whole host of positions will be unavailable because of the *illusio* of the field. It may therefore seem a reasonable and rational strategy to undermine your opponents' position as guardians by attacking mosque buildings, despite the fact that the underlying driving force in the conflict may not necessarily be ethnic or religious. Attacks on religious buildings are in this instance only a consequence of the structure of the field. By regarding the destruction of religious buildings in this light, this action acquires logic distant from the havoc of ethnic cleansing.

The accumulation of capital is a principal activity on the field. However, the conflict on the social field also has another level: the actual definition of what is to be regarded as legitimate capital, and therefore, ultimately, the definition of the field itself is also an object of contention within the conflict: the definition of the field is always in contention and is an important point to appreciate.[13] A basic example might involve deciding to what degree physical violence has legitimacy as a political means on the field at a given time. All sections of the intervention force, military as well as others, will work towards the same goal, a comprehensive approach, which in this case can be defining the use of violence as an inappropriate strategy as a means of achieving or exercising political power. A militarily strong minority section of the population will see obvious disadvantages in the democratisation of their society, since their percentage part of the population does not equate to their military strength. Representatives of such a group may choose a delaying strategy against the development of democracy, in order to convert their military power base into a form of capital more marketable in the future. Actors, who choose not to forego violence as a political means, will find that their power will stagnate as the use of violence is limited by intervention

---

[13] Bourdieu (1996b, p. 44).

force operations. At the same time, all the groups who have rejected violence will receive strong backing from the resources that can be generated by both the military force and organisations cooperating with them. In this situation, soundings will be conducted and proposals to forego violence made to the groups that persist with a violent strategy.

BIBLIOGRAPHY

Ångström, J. (2003). Concepts Galore! Theory and doctrine in the discursive history of low intensity conflicts. In J. Ångström & I. Duyvesteyn (Eds.), *The nature of Modern War: Clausewitz and His critics revisited*. Stockholm: Swedish National Defence University.

Bourdieu, P. (1992). *Texter om de intellektuella. En antologi*. Eslöv: Symposion.

Bourdieu, P. (1996a). *The rules of art*. Cambridge: Cambridge University Press.

Bourdieu, P. (1996b). *Homo academicus*. Eslöv: Symposion.

Broady, D. (1989). *Kapital, habitus, fält. Några nyckelbegrepp i Pierre Bourdieus sociologi*. Stockholm: Universitets- och högskoleämbetet.

Gunneriusson, H. (2002a). *Det historiska fältet. Svensk historievetenskap från 1920-tal till 1957*. Uppsala: Studia Historica Upsaliensia.

Gunneriusson, H. (2002b). Sociala Nätverk och fält, så förhåller de sig till varandra. In H. Gunneriusson (Ed.), *Sociala nätverk och fält*. Uppsala: Opuscula.

Guttenplan D. Don. (2005). *Förintelsen inför rätta. Sanning, lögn och historia: berättelsen om David Irving-rättegången*. Stockholm: Norstedt.

Schelling, T. (1996). *Arms and influence*. New Haven: Yale University Press.

Sun Tzu. (1997). *The art of war*. North Clarendon: Tuttle Publishing.

Weber, M. (1989). *Ekonomi och samhälle*. Band I. Lund: Argos Förlag.

CHAPTER 3

The Political Field in the Operational Area

Abstract This chapter deals with one possible implementation of field theory in a generic operational area. The main challenge is to operationalise the concept of field-specific capital in order to classify and affect the adversary and other actors on the field.

Keywords Operationalisation · Field specific · Actors

The classic Western definition of war reads: "the continuation of politics by other means".[1] This definition should be kept in mind when dealing with the armed groups in the area, as well as the civilian political structure, regardless of the form or level of violence prevalent on the field at the time. If the local field rules and recognised forms of capital can be identified, then the chances of being able to make effective decisions will increase dramatically.

For example, patterns of behaviour can be monitored to predict future activity at a strategic level and below. These predictions gain further credence when the intervention forces actively use their knowledge to bring about structural change to the logic being applied in the operations area; it is not just a question of observing. It is pertinent here to

---

[1] von Clausewitz (1991, p. 42).

© The Author(s) 2017
H. Gunneriusson, *Bordieuan Field Theory as an Instrument for Military Operational Analysis*, New Security Challenges, DOI 10.1007/978-3-319-65352-5_3

remember that this is an empirical science. The theoretical framework given presents a method of reflecting on and an approach to the mission in hand. What in reality actually applies and has import on the field is beyond the scope of this theory; the theory must be developed heuristically. This can be achieved by conducting various types of InfOps (Information Operations), such as HUMINT (Human Intelligence). Strategies on the field are all a type of reproduction strategy, the essential being to maintain or improve one's own or the group's position on the field. In short, it is a question of power. Conversion strategies aimed at changing one form of capital to another, which in a given situation appears more advantageous, are one example.

In an area plagued by unrest, the political field may be likened to a piece of sloping ground where actors believe that they must use violence to have any influence, thereby foregoing more civilised political methods. In general terms, one can say that the ultimate goal in the struggle on the field is *dominance*. *Dominance* is achieved by the actors who have acquired a considerable amount of the current *marketable* symbolic capital on the field (e.g. capacity for violence, ability to call mass meetings and religious legitimacy), often manifested by occupying important positions (e.g. as leader of a particular group or organisation). The shape of the field will reflect the values of the dominant group or groups. A field presupposes a conflict, which then defines the arena. All actors believe the political game to be worth playing. This is a prerequisite to qualify as an actor, but the actors hold different viewpoints and apply different methods. The presence of a number of institutions that can dispense awards within the field raises the stakes for conflict on the field. These may be state institutions, but they may also be informal institutions such as various groupings (military, ethnic, religious, geographic, etc.).

The international force will have as a goal the establishment of a monopoly of the use of force in the operations area. This will mean that several of the actors on the field will have their positions threatened. To merely meet these actors with force will compel them to go underground, which is not a step towards a desired end state. The desired effect in the local political field of the operations area is that all political activity is conducted within the framework defined by the politico-strategic goals of which the intervention force is a product. Groups which have an ulterior political agenda, beyond violence, will be inclined to reappraise their strategies if the alternative is marginalisation, or the threat of their organisation ceasing to exist. Marginalisation in this sense

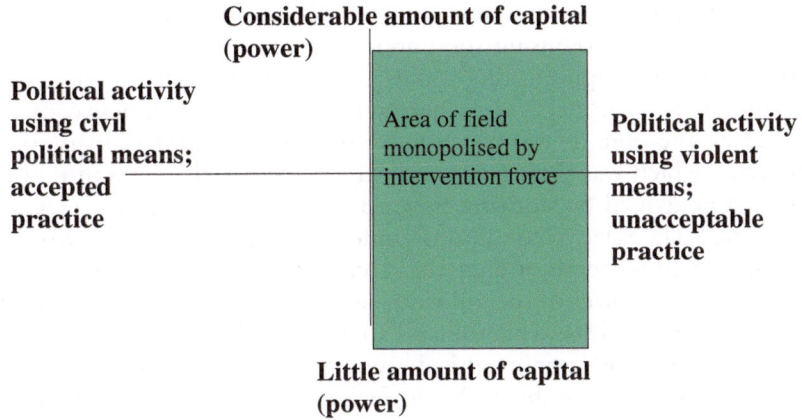

Fig. 3.1 A field sketch used as a mind map. Actors on the *right* side of the field will, by the use of both the carrot and the stick, be made to change their strategies to ones more akin to those on the *left* side of the field

is primarily political, but social or economical marginalisation can be critical vulnerabilities to weaken their political capital. This inclination is one of the two crucial factors which make it important to influence these actors who can choose in particular. The other important factor is that actors with a political agenda find it easy to assert legitimacy for their actions, in contrast to the bandit gangs and warlords with limited political aims and little choice other than marginalisation (Fig. 3.1).

Different types of groupings can offer various types of reward, both formal and informal. For example, money or a promise of a position in the present or future government hierarchy may be offered. However, some driving forces may be less easily identifiable. As a rule, driving forces are harder to identify if they involve cultural phenomena that are not easily recognisable within one's own political culture. It is in these obscure peculiarities that there lies a potential source of great error with regard to one's choice of course of action in the operations area. Power in one form or another is what is respected on the field, even if that power is of an indirect nature. The specific initial empirical question in any study is: What constitutes power in this situation?

Each field has its own specific characteristics, even if all fields have certain common basic structures that enable field theory to be applied in different areas.[2] For example, the French intellectual field cannot be studied without recognising the importance afforded to a few so-called elite schools throughout the country. A country such as the former Yugoslavia had the particular characteristic that the distinctive quality of each of the regions was much more pronounced than in many other countries, which was then reflected in the political field. A picture of the field can be gained by studying how actors and institutions interact with each other. The relative conditions guided the actors' strategies, which in turn served to preserve or alter the strength of the field's various forms of power.[3] The power of the individual actor on the field is relative to the remainder of the field, which means that a stronger position is attained if the capital of one's opponent is devalued, and his power thus reduced.

The model of thought is simple in itself, but it is important to do more than just understand it. It should be seen as an approach or link to reality that quickly enables one to organise and structure the actors in an operations area. As a result, the model offers a guide to the courses of action open and to what may be appropriate in any given situation. Acting in accordance with field theory at the operational level requires one to focus on three main tasks, of which one can be considered the one that distinguishes the method of attack from other more conventional alternatives.

A Tentative Checklist

Initially one can say that the checklist should be seen as focused on structural change first. This will give collateral effects on the actors which are directly affected by the structural change and secondary effects on the actors, who are not directly involved in the practice being targeted, but still are a part of the social field and takes notes on what happens on it.

1. Secure a monopoly on armed force

This is the minimum and basic demand placed on the intervention force. If the politico-strategic goals are reasonably modest, then the area of operations may be limited to an area smaller than the whole conflict arena. Bosnia provides a good example, when for a long time the

---

[2] For several examples of field studies, see Broady (1998).

[3] Bourdieu (1996, p. 159).

Bosnian Serbs were allowed to control their own areas because there were insufficient resources to contain their capacity for violence.

2a. Establish an understanding of the logic of practice in the operations area

This is the structural approach, which deals with making out what is considered of value on this social field, what is generating capital, both symbolic and economical capital, perhaps even social capital.

2b. Establish the agendas of key actors

This applies in particular to those who perpetrate violence: Which of them have political goals which can be achieved by means other than violence? Those that have this type of goal are potentially able to change their approach to a political agenda without violence. In addition, those actors who have a political agenda but who do not use violent means must be identified. All of these actors must be accessible for dialogue and qualitative intelligence work.

3. Work on what are considered the critical points of the logic of practice

This is most likely a heuristic process, where the targets and goals change during the process as new learnings most likely will submerge the more you deal with the social field of the operational field. One example of what one can target is the way religion is used as a means of mobilisation of violence. Another example can be the informal economy of the social field, how to tweak it to get the actors into perceiving other (more benevolent) strategies as "the best" strategies, or at least better than the strategies their own forces regard as unwanted. Furthermore, the symbolic capital of social, geographical or ethnic groups, even gender, can be can be targeted—strengthening or weakening their positions in order to achieve certain effects on the way the social field is constituted.

4a. Offer actors using violent tactics the opportunity to change strategy

Once an acceptable monopoly on armed force has been established and has been in operation for a while, negotiations with the actors should begin. In short, resources are offered to allow participation in the political process, but using peaceful means. The choice of resources will vary and will be guided by the actors' needs, and may involve both material and services. Reconstruction is thus not something that takes place after the conflict. Reconstruction of whatever is involved provides a route to gaining control over the area of operations; it is not something that happens through the process of establishing control. The military force must therefore direct the process of reconstruction in the area, partly because it is an integral part of their operational plan as discussed here, and partly because it would be a dangerous undertaking for NGOs to operate without the protection of the military force.

4b. Offer resources to political actors not using violent tactics

There should be a carrot for political actors with an existing peaceful agenda. It is, however, just as important to have a stick ready for the actors using violent methods who refuse to change their strategy. They will see their political influence wane at the same time that the political party apparatus of politicians using peaceful means and its ability to reach out will grow, and that the military force continues to hinder the agenda of violent political elements. The offer at (3a) remains open for those who wish to take it up. It is a question of showing which practices will lead to increased power, violence probably having been the most advantageous prior to the military force initiating their operational plan.

It is important at this point to stress that those who receive political help are at liberty to disagree with the politics pursued by the countries represented within the military force. Under no circumstances should it appear that those receiving help have been bought up by a foreign power. Their symbolic capital should not be undermined, which will be the case if those receiving support are perceived as nothing but puppets of the contributing nations of the military force. The ideal situation would be to have the military force supporting a group comprising different collective actors covering a wide political spectrum. These actors would then become the foundation for the remodelled political field. Through the strengthening of these actors, the political field will be redefined, leading to the actors inclined to violence being marginalised. In the end, if these actors do not change strategy, they will be perceived—not only by the international force but also more importantly by the population—as terrorist organisations opposing the civil political system.

BIBLIOGRAPHY

Bourdieu, P. (1996). *Homo academicus*. Eslöv: Symposion.
Broady, D. (1998). *Kulturens fält. En antologi*. Göteborg: Daidalos.
von Clausewitz, C. (1991). *On war*. http://www.gutenberg.org/files/1946/1946-h/1946-h.htm [Visited 170320].

CHAPTER 4

Field Operations as Information Warfare and Operationalisation of Theory

Abstract This chapter touches on concepts of information warfare, for example COIN-theory. Field theory as information warfare is the main topic. The purpose is to frame the field theory as a valid example of both method and theory regarding information warfare. This is especially important as a lot of the practice regarding information warfare is not anchored in a social theory but mostly in a best practice way of acting. That is all well but a theory provides a wider range of generalising and thus options of free-thinking.

Keywords Information warfare · COIN · Field theory

InfOps are mainly associated with guarding access to one's own information, but InfOps are also concerned with controlling the opposition's access to information.[1] It is essentially the process of disinformation or denying information. The term InfOps is comparatively wide ranging and can involve operations that are not necessarily conducted during times of crisis or war. Information warfare is conducted with the aim of influencing an area suffering from war, crisis or other violent circumstances. Can the application of field theory therefore be said to constitute information operations? It must be admitted that it is an unorthodox

---

[1] *Dictionary of Military Terms* (1999), London.

© The Author(s) 2017
H. Gunneriusson, *Bordieuan Field Theory as an Instrument for Military Operational Analysis*, New Security Challenges, DOI 10.1007/978-3-319-65352-5_4

form of warfare. The answer is affirmative, with the finesse that the enemy does not need to be completely identified but just by his methods and goals. The aim is to attack a critical vulnerability of the enemy by altering the social, political and economic reference system. In a paraphrase of a famous saying by Mao, one might say that it is a question of changing the water in which the fish are swimming. The critical vulnerability of the enemy is his political legitimacy in the mission area. The attack is conducted by changing the conditions of the field so that even unidentified political actors may also find themselves under attack. If the vulnerable point is attacked, the enemy's centre of gravity will have to be adjusted accordingly. For this approach to be effective, the actor must be a genuine political actor, an actor interested in political gain. If the actor has no political ambitions—perhaps in the case of organised crime for financial gain—then there is also no reason *not* to countermine the actor's position as a player on the field. However, it is important that there is a general perception that the actor is not a political actor. It is the judgement of the field, not that of the actor, that determines whether an actor is political or not.[2] It is worth stressing the importance of qualitative knowledge of the actors on the field, in the same way that knowledge of the field's own structure is key information.

A field theoretical approach to operations should strive to give the enemy information of a type that is not disinformative. As it is a long process, it is important to not undermine the trust in the process with disinformation. The information presented should fit with the stated aim of influencing the actor to change his strategy. It involves information in the form of action rather than words, to clearly show the disadvantages of a strategy of violence by confronting him with a well-armed and well-equipped military force. It is also important to clearly show the alternative strategies that are possible. The latter should be combined with making it obvious that the positions of non-violent actors will be greatly enhanced, whatever their political standpoint. In this way, it will become clear to all that those with a non-violent political agenda will experience a considerable enhancement of their power, while the increase in power for the actors using violent means will be obstructed or will suffer a reverse. It is thus important to inform the enemy in the correct manner. As this will change the distribution of power on the field, one might see different and new

---

[2] However, the field can be influenced to form a certain opinion of a particular actor.

avenues of action being taken which can only be determined by empirical observation rather than theoretical laws. One can, for example, see that factions try to push a more civil branch of action while still not completely abandoning the capacity to violence in another respect, in effect a possible case of lip service. One might also see that complicity might result in armed groups of a faction (whatever its constitution) might turn on what they perceive as collaborators. In the latter case, the capacity of violence by our own forces must not be dismissed as unnecessary.

Nevertheless, field theory information efforts have clear strategic implications, at least if by strategy one means creating an overall plan and materially providing sufficiently for the operational level to ensure that information can be communicated. It will require information and reconnaissance resources over and above normal levels if a field theory approach is to be more than just a shot in the dark. Information has been stated to be the key to irregular warfare, something which hardly can be denied.[3] One can because of that clearly see the advantages of using a perspective as Field theory—which bases its foundation on information—as an operational COIN-approach.

Political activity is a manifestation of the will to acquire power through the accumulation of capital in the political field. By reading off the rules of the field, a range of measures can be instigated against various actors on the field to influence them to more or less voluntarily change their operational strategy. How this will be achieved is an empirical question and one that does not lend itself to theorising; all fields are unique and must be treated accordingly. It is a question partly of the interests and competence of the actors, and partly of the range of resources at the disposal of the international force. The *DIME* approach is one of many methods used to structure the various arenas used for an operation.[4] This model can be useful in certain situations, but is far too general in most cases to be of interest for our purposes here. There are a number of possible measures of symbolic nature that actors in the political field in the operations may find attractive. Provided that the actors (collectively or individually) forego the use of violence as a political means, a range of offers can be made in return. These offers must be made clearly and abundantly to the actors

---

[3] Kalyvas (2009, p. 174), also quoting Eckstein (1965, p. 158).

[4] The acronym has later expanded to DIMEFIL (Diplomatic, Information, Military, Economic, Financial, Intelligence and Law Enforcement). Headquarters Department of the Army, p. 1.

who are already positioned on the left side of the field. They must at no time be seen as a reward to the actors on the right side of the field. The maintenance of a strategy of violence on the right side of the field must be met with one of force until the strategy is clearly abandoned in favour of one that fits on the left side of the field. Only using force is the simple coercive approach. True, it is part of the structuring of the social field, but more restructuring of the field must be made—open up new attractive avenues of action for both actors who are benevolent but also others. It is important that the opportunity to switch strategy is always available and is actively encouraged.[5]

Examples of incentives to change strategy:

- Positions in public administration—dependent on competence and interest. Make them dependent.
- Consultation with actors on legislation—in line with a stated political agenda. Draw them in.
- Building up infrastructure—investment, priority, naming of installations (e.g. bridges, road systems and infrastructure for telecommunications). Hearts and minds.
- Building and siting of schools—for example specific home areas can be given priority. One can discredit as well as credit with tactical help.
- Alternative job opportunities—guaranteed employment for former militia. If Maslow's pyramid of needs is important in the case, then this will have effect.

The above are but a few examples of the incentives that might be offered to get actors to change their strategies, it is no way everything that can be done. It is important to read the actors and their priorities correctly; they maybe have no interest at all in job opportunities, but for example may well have an agenda for legislation. For instance, proposed legislation may be submitted to them for their consideration or they may even be

---

[5]The action taken by the USA against Iraq's Sunni leaders may be seen as a variation on this theme. They can always participate and influence the political process—influence developments—if they want to. The problem is that they will never acquire any real power when other ethnic groups make up 80% of the population. In this particular situation, the USA should have found alternative solutions to the problem by making the Sunni Muslims feel that they were part of the process.

asked to take part in dialogue for the formulating of that legislation. For other groups, the most important thing may be securing an income for their members under safer conditions; this may be all the incentive necessary. Furthermore, many projects bear symbolic values far beyond the practical value the project brings. "Roads ain't roads", as David Kilcullen expresses it.[6] They can signal an ambition of increased presence, military or other, it signals long-term commitment and ambitions.

The field theory approach aims at getting the actors to adopt strategies in line with the conditions that one has set for the field; they must be made to see that actors with non-violent strategies quickly acquire power and influence in society—in contrast to those who espouse violence. It is important to work to suppress violence and *simultaneously* actively promote the actors who support a non-violent political process. Doing this will tilt the power balance on the political social field in a direction wanted. The tools available for promoting these actors may come from within one's own organisation or from other organisations. Non-profit organisations are also actors on the field, regardless of whether or not they are working alongside the military force. The problem is that independent organisations have their own agendas different from that of the military force. It is also often the case that an NGO will lose respect in the area of operations if it works with the military force. It is often sufficient for NGO representatives to merely be afraid that this will happen, for cooperation to worsen. This, however, is something that needs to be addressed from case to case as it is an entirely empirical question.

Quantitative basic facts, used for positioning the actors, can over time be complemented with data that may not appear to have anything to do with power on the field, facts that may seem trivial. For example, details of choice of brand of cigarettes may be of interest. If it is seen that certain brands are consumed in great quantities or not at all by certain groups without any reasonable explanation, then one might have found something interesting. If a particular group smokes only a certain brand of cigarette this may point to smuggling activity which in its turn drives the informal economy.

---

[6] Kilcullen (2009, p. 108).

How can field theory be operationalised? How does one decide where actors are positioned within the field? There is a choice of the two general methods of analysis, either qualitative or quantitative. First is something about the quantitative method. A special software application programme has been produced for quantitative analysis, and it is well known among those who make correspondence analyses. The method has been mentioned above and is called *multiple correspondence analysis* (MCA). In short, the process involves gathering a large amount of empirical data, regardless of whether or not it initially appears useful for defining the actors on the field. The information may include patterns of behaviour, social networks, consumer habits, economy, local political issues, what the different parts of society are focused on, etc. The results of this information gathering will then be processed to provide an empirical definition of the groups. Certain values or habits for the various actors may come to light that have not previously been noticed. Some information may be of the type that affords the actors prestige, respect, the opportunity to exercise power, access to various arenas such as the media, religious circles. All these factors are then processed, and an overall positioning of the actors within the field is then produced. An actor's position on the horizontal axis will be wholly dependent on his readiness or otherwise to use violence.

The second method involves a qualitative approach. As a result of observations made, the various actors are then positioned where one believes they fit on the field. This method may appear less reliable, but fulfils its purpose. The quantitative method relies on adopting the right parameters; the qualitative picture is based on an overall impression gained by studying and interacting on the field.[7] What is appropriate is to a high degree dependent on a qualitative consideration of the analysis results. The quantitative method can be used at the strategic level where it is easier to gain an overview, because there often is a larger amount of incoming data than one can deal with. At the strategic level, the analysis can be used to accurately gauge which resources will be required at the operational level to enable successful accomplishment of the tasks identified by field theory analysis. The qualitative method can be used to advantage at the operational level, where one is better able to feel the

---

[7]Correspondence analysis assumes a certain familiarity with the use of software, but courses are available: http://www.skeptron.ilu.uu.se/broady/sec/k-kor04.htm [Visited 170509].

pulse of the field on the ground—it is at this level that field theory has most potential. In the end, the need to gather information will lead to a demand for certain capabilities at the tactical level, even if the analysis is not conducted at that level. With the quantitative method, one gets a numerical value, with decimals. There is no reason to believe that the quantitative result is more valid than a qualitative one, as there are often qualitative decisions behind the quantitative ones. For example, what should an attack constitute to generate a number in the matrix and so on. With the qualitative method, you get an analysis which you have to decide on in terms of yes/no, do/don't. The lack of numbers might superficially seem a less reliable method, but it is all determined by the reliability of those doing the research, not the method itself. A decision about a yes or a no is often what you need in a given situation, no more, no less. Therefore, the qualitative method is sufficient as a method for decision-making.

Field theory sets great demands on intelligence work, but at the same time, it has been made easier with the advent of new technology. Field theory will carry with it its own genre of HUMINT.[8] Something which is not often mentioned is that the level of training for the military has been improved over the last few decades. This does not only apply to officers, but also to soldiers who have civilian professional qualifications.[9] One has to understand that the intelligence community is far wider than military intelligence, and one has to use all sources for intelligence there are. In quantitative terms, the amount of information available has increased over time, to a large extent made possible by technology. It is important at this point to clearly differentiate between the task of merely identifying targets and qualitative intelligence work.[10] For this reason, scientific methods and theories must be used to a much greater degree

---

[8] HUMINT can mean different things in different contexts, see Ferris (2004, pp. 59 and 67) (Intelligence Operations).

[9] Ample opportunity exists to raise the level of education for soldiers at all levels in the future, by integrating university education (in this case in municipal adult education). This author will present thoughts on this subject on another occasion.

[10] Ferris (2004, p. 57). The difference between target identification and intelligence is indicated, but not the need to define which agencies are involved.

than before—it is no longer enough to just gather all relevant information and start off. All relevant information will amount to too much for it to be managed without using a feasible method and approach.[11]

One problem concerning method is being able to differentiate between actors employing violence who have a political agenda, from those whose use of violence serves no political purpose at all. It is important not to ignore the requirement to address the problems posed by the criminal elements without political agendas, but one must quickly understand that they are not interested in exchanging their violent tactics for mass meetings and printing presses—they are not interested in any peaceful political strategy. The problem is illustrated by the following example from Iraq in summer 2005:

> The shootings became so frequent in Baghdad this summer that Horst [Brig. Gen. Stf. C. 3 Inf. Div.] started keeping his own count in a white spiral notebook he uses to record daily events. Between May and July, he said, he tracked at least a dozen shootings of civilians by contractors, in which six Iraqis were killed and three wounded. The bloodiest case came on May 12 in the neighborhood of New Baghdad. A contractor opened fire on an approaching car, which then veered into a crowd. Two days after the incident, American soldiers patrolling the same block were attacked with a roadside bomb. [- - -] Horst declined to provide the name of the contractors whose employees were involved in the 12 shootings he documented in the Baghdad area. But he left no doubt that he believed the May 12 incident, in which three people were killed, led directly to the attack on his soldiers that came days later on the same block. "Do you think that's an insurgent action? Hell no," Horst said. "That's someone paying us back because their people got killed. And we had absolutely nothing to do with it.[12]

---

[11] Only recently has it been axiomatically maintained that all relevant and reliable source information should be used in historical research, and something which was established in the earlier part of the 1900s in Sweden by Professor Lauritz Weibull, who was particularly interested in method. He was, however, primarily interested in the Swedish Middle Ages where too much information was never a problem. Researchers who are interested in current times must learn to filter information in a systematic way.

[12] http://www.washingtonpost.com/wp-dyn/content/article/2005/09/09/AR2005090902136.html [Visited 170509].

The quote illustrates several relevant factors, one of which is that there is a requirement for observers on the ground to identify who is who on the field. It requires an almost hermeneutic method of observation, which in turn means that observers must have a good pre-understanding of whether an observation will lead to any relevant conclusions. In addition, the quote shows that it is not always easy to tell the difference between actors with a political agenda from those who merely react in trigger response in revenge for an incident. However, that it is difficult to differentiate between the two does not under any circumstances mean that one should cease to maintain an analytical approach. Finally, the example shows what happens if one does not know one's own operational aims. The antithesis of field theory is to internalise an approach in which one indiscriminately escalates the spiral of violence; achieving peace in the foreseeable future will be quite out of the question with the approach outlined above. In addition, even if the insurgents can see the differences between different Americans in this case, they might refuse to act correspondingly.

> This mind map of a social field (do notice the characteristic 4-fields) focus on the right part of the field (in the oval). It focuses on those who conduct politics with violent methods for different reasons. Many of the groups there are dependent on the rules of other fields than the political one (primarily the economic field).

The diagram above shows the need for a deeper identification process after one has identified the actors using violent tactics. The question to answer after initial identification is the ever relevant *why*. Why do they have an agenda that involves violence? Only after these why-questions have been answered will it be possible to formulate a practical, purposeful course of action to engage these actors—and also those who already have an acceptable agenda. The engagement can be directly on the actors or towards structures on the social field which one estimates would affect these actors. *Once again, it is more rewarding to act on changing the social field than aiming at certain actors directly, as it is not sure or even likely that you have identified all the actors you want to target.* It is important to remember this, even if this discussion to a large extent deals with the actors. Aiming at the social structures will in its turn affect even actors you still don't know about. In the end, it is humans and their behaviour you want to see a change in, but the road might have to go through structural change first.

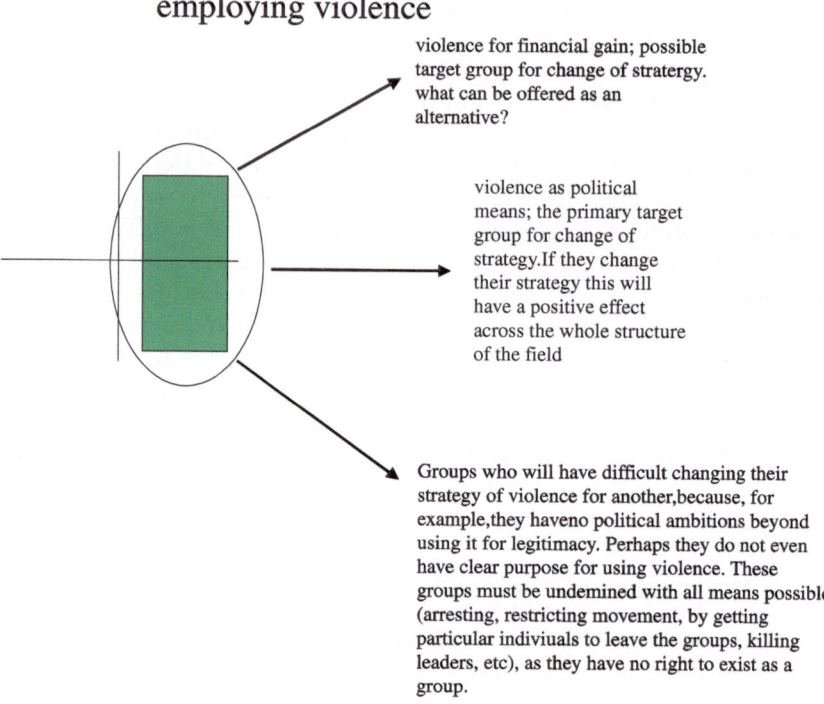

employing violence

violence for financial gain; possible target group for change of stratergy. what can be offered as an alternative?

violence as political means; the primary target group for change of strategy.If they change their strategy this will have a positive effect across the whole structure of the field

Groups who will have difficult changing their strategy of violence for another,because, for example,they haveno political ambitions beyond using it for legitimacy. Perhaps they do not even have clear purpose for using violence. These groups must be undemined with all means possible (arresting, restricting movement, by getting particular indiviuals to leave the groups, killing leaders, etc), as they have no right to exist as a group.

Fig. 4.1 Mind map dealing with violent actors

All actors, certainly not the collective ones, can easily be divided into just one of the social categories constructed (e.g. as in Fig. 4.1). One can imagine a group of people who had all three categories shown above represented organically. But as with all theory, the analyst must draw a good-enough line somewhere. As it might be a matter of life and death in this case, there is a need to communicate the perception of the actors to the actors, if one wants to affect their logic of practice.

Purposeful action means the military force adopts a course of action aimed at getting the actors to voluntarily change their strategy, involving methods beyond the purely firefighting techniques employed to prevent genocide and the like. A group that has no political agenda linked to its use of violence is difficult to influence using methods aimed at getting them to change from their current pattern of behaviour to acting peacefully in the political arena. In this case, it might involve actors who exercise force in the operations area and who thereby make financial gain through smuggling. The exercise of force is an outward manifestation of maintaining control over a smuggling operation. These actors are more interested in economic power than political power and should be tackled accordingly. This knowledge is of course important to enable an understanding of the actors and motives driving the current wave of violence in the operations area. David Kilcullen writes that in Afghanistan 3000–4000 of the Taliban (which were approximately 10% of them) were "hard-core fanatics who are not reconcilable under any circumstances".[13] It is important to use not only physical violence, but foremost symbolic violence on these individuals. This can be discrediting them in religious, social, political, ethnic or economic ways. It is either this or giving up the idea that there is a cultural element in conflict, which there often is. That we, the West, for example, want to impose democracy on "the others" as that is something the West as a society believes to be objectively right like a jihadist believes in Islamic law as objectively right. Some people are hard to reach and that must be acknowledged as a fact and treated accordingly, with force if necessary.

Another method problem is the purely operational one of establishing collaboration between the military units and the various forms of voluntary organisations (NGOs). The latter often have a policy not to collaborate with military units. However, this problem lies outside the scope of the possible for a study of this nature; it will require a favourably pragmatic approach to operations in the field.

By way of introduction, it should be said that ethnic cleansing and genocide do not occur in all conflict situations. No matter what International law says; there are plenty of examples of people being forced to flee an area, or being killed just because of ethnicity and not because of some kind of actions taken on the victim's behalf. There is no shortage of plight and suffering among those who find themselves with the "wrong" ethnicity or nationality. This text is concerned with operational realities for forces on the ground, rather than definitions made for political reasons. International law is only normative if there are interests from important national states—which often can be condensed down to the USA, Russia or ex-colonial states in the region—so if an event

---

[13] Kilcullen (2009, p. 49).

is called ethnic cleansing or not by International law is to a large extent politics. This does of course not stop experts in International law from having opinions that differ from what the world community decides is ethnic cleansing or not. At least ethnic cleansing has, however, proved to be a common element of the low-intensity conflicts waged since the Cold War. The process follows a relatively linear course and the breaking point may be said to illustrate a case of "external shock".[14] It is then that the legitimacy of current norms is questioned in practice and not just in rhetoric. The reasoning also pre-supposes that explanatory models, such as Christopher Browning's in *Ordinary Men: Reserve Police Battalion 101 and the Final Solution in Poland*, are valid. Namely, that perceived conditions are of decisive importance, if not of sole importance, as a precursor to genocide.[15] In the case of Yugoslavia, the process took place sequentially in a number of stages seen as separate outbreaks of the fighting. The fighting in Eastern Slavonia is an example of comprehensive state-controlled ethnic cleansing. This same type of phase occurred in the opening stages in Bosnia when the Bosnian Serbs enjoyed their initial successes. The expelling of, for example, the Kosovo Albanians from Kosovo may be seen as another key phase in the complicated course of events in Yugoslavia. This purge was conducted by special units, like *Arkan's Tigers*—led by Željko "Arkan" Ražnatović—who initially demonstrated a marked capacity for violent and spectacular action.[16] Some think that it took until 2002 for the logic of ethnic cleansing just to begin to erode, but still not to be reversed.[17]

When the method for analysis has been chosen—quantitative or qualitative as discussed above—the next question to address is which method will be used to start influencing the field. In short, it would be fair to say that the method chosen will be a mix of the alternatives positioned between *targeting* and *hearts and minds*. The former involves getting to the actors one wants to influence, by either providing incentives or

---

[14]Farrell (2005, p. 14).

[15]Browning (1998, p. 173).

[16]The fact the Arkan's Tigers had a film team with them on at least one occasion strengthens the theory that their mission was to create conditions for a war of ethnic character, de Graaff (2003, p. 118). See also Allin (2002, p. 59).

[17]Allin (2002, p. 44). See Sell (2002). for more about paramilitary forces at the opening stages of the war, p. 165.

limiting their room for manoeuvre (which in extreme cases may mean killing someone, a fact one should not turn a blind eye to when military operations are involved). The latter involves altering the structure of the field so that certain patterns of behaviour become unacceptable in the social field of the area of operations. In reality, it rarely involves a narrow focus on either actors or structures, but a blend of alternative courses of action.

As an epilogue, it is fair to mention that some actors are multifacetted and might have different appearances at the same time. One has to choose how to act with caution. An actor like PIRA, for example, could choose to use Sinn Fein, the political arm of the movement if the military tactics were less suitable at a certain time, only to shift to the military focus later on. The same goes with Hamas and Hezbollah, for example, they too can act on different parts of the social field depending on the circumstances. This does not make the model less suitable, it is very suitable in order to identify how these actors behave and why.

BIBLIOGRAPHY

Allin, D. H. (2002). *Nato's Balkan interventions*. Oxford: Routledge.

Bowyer, R. (1999). *Dictionary of military terms*. London: Bloomsbury Reference.

Browning, C. R. (1998). *Helt vanliga män: Reservpolisbataljon 101 och den slutliga lösningen i Polen*. Stockholm: Norstedts.

Eckstein, H. (1965). On the etiology of internal wars. *History and Theory, 4*, 2.

Farrell, T. (2005). *The norms of war: Cultural belief and modern conflict*. Boulder: Lynne Rienner.

Ferris, J. (2004). Netcentric warfare, C4ISR and information operations. In L. V. Scott & P. D. Jackson (Eds.), *Understanding intelligence in the twenty-first century. journeys in shadows*. London: Routledge.

Graaff de, B. (2003). The wars in former Yugoslavia in the 1990s: Bringing the state back in. In J. Ångström & I. Duyvesteyn (Eds.), *The nature of Modern War: Clausewitz and His critics revisited*. Stockholm: Swedish National Defence College.

Kalyvas, S. N. (2009). *The logic of violence in Civil War*. Cambridge: Cambridge University Press.

Kilcullen, D. (2009). *The accidental guerrilla. Fighting small wars in the midst of a big one*. Oxford: Oxford University Press.

Sell, L. (2002). *Slobodan Milošević and the destruction of Yugoslavia*. Durham: Duke University Press.

PART II

Empirical Example

CHAPTER 5

The Pre-war Strategic Situation in the Balkans from a Field Theory Perspective

Abstract The following is a short account of the events leading up to war in the Balkans, in particular in Bosnia. The intention and over-all description in these few pages will help provide an example of the practical application of a *field theory perspective* to conflict, rather than describing the conflict in full. Some actors are chosen as examples for the purpose of theoretical explanation, even though there are others to focus on if one wants a complete historical account. Main actors as for example Slobodan Milošević and Franjo Tudjman are also presented in this chapter.

Keywords Balkans · Bosnia · Slobodan Milošević · Franjo Tudjman

Something which must be stressed in this discussion is that Bosnia is a relatively small country (51,129 square kilometre), about the size of Costa Rica.[1] The limited area is important to note. It therefore requires very few criminal elements to influence opinion to the point where peo-ple are prepared to carry out ethnic cleansing by directive or as a form of revenge. A system of informal criminal networks, involved in the

---

[1] *CIA Factbook* (2017). https://www.cia.gov/library/publications/the-world-factbook/ [Visited 170509].

© The Author(s) 2017
H. Gunneriusson, *Bordieuan Field Theory as an Instrument for Military Operational Analysis*, New Security Challenges, DOI 10.1007/978-3-319-65352-5_5

trafficking of people, weapons, drugs etc., was established as in all conflicts, with various actors using violent means to establish their territories (meaning here primarily not physical territory).[2] Partly, because of this professor, Mary Kaldor has described this form of conflict and this specific conflict as a new form of war. Kaldor thus writes that she has identified a new form of war which has emerged since the Cold War, involving more or less organised violence between different parties (often not states). The motives behind this phenomenon are many and varied.

As I put a theoretical perspective of my own on the conflict, I would like at this point to dwell a little on Kaldor's ideas, ideas which have met some criticism.[3] Her view of the informal economy as the driving force behind the war is essentially correct; she makes a fundamental point when highlighting the presence of the criminal world as an influential force. In bourdieuan terms, one could refer to a changing social field. That said, I would like to take issue with many of the theories she promotes. One problem is that history shows us plenty of examples of the fact that the parameters for the "new" wars already have existed and still exist; it is easy to recognise the phenomenon from history. The informal economy is not new, certainly not where war is being waged. People in the West today are used to regarding war as something that takes place between sovereign states. However, one does not need to look too far back in time to see the processes that Kaldor describes as part of just about every nation-building process that has occurred in Europe (this is probably also true for other parts of the world, but certainly holds generally throughout Europe). She highlights globalisation as a new contributing factor to this new type of war. It must be said that the consequences of war now have a more wide-ranging influence, but that this should affect events in a qualitatively new fashion remains to be proved. Breeding grounds for unrest have taken on a new significance within criminal circles, when all sorts of illegal transport make its way through the lawless country like electricity conducted through copper. There has,

---

[2] For an interpretation see Kaldor (1999, p. 105).

[3] Krampe (2010).

however, always been interest shown by criminal groups to fill the vacuum left by a (failed) state; therein lies nothing new.[4]

Kaldor focuses often on the Napoleonic Wars, which were essentially state-controlled, and consequently her studies contain a large number of clausewitzian references. She also focused on the state entity in his analyses, for which he in turn received a fair amount of criticism.[5] What also occurred during the Napoleonic Wars was that Westphalian Europe, in many regards what remained of the Europe of the Middle Ages, was struck from the map. Territorially cohesive areas came to dominate, ruled over by a power often concentrated in one person, such as a king or emperor.[6] The period clearly encompasses the little-discussed history of the losers of the time, revealing the struggle of the small nations against the large national states, and this applies equally to the course of events that not only Kaldor described as state-directed war. There is therefore cause to exercise caution when one generalises, because contradictions can remain hidden in examples chosen. However, to speak in qualitative terms of new forms of war is just as unwise as stating that revolutions occur during war.[7] It is certainly true that Kaldor makes many commendable points in her book, they predominate without question. What I do take objection to is the intractable desire to discover something new when there exists little foundation for the theory. Talking about a new type of war is going too far, because the distinction made hardly warrants such a conclusion. On the other hand, Kaldor presents a reasonably adequate picture of how violent conflict, with states as just one of many actors on the stage, has manifested and continues to manifest itself.

---

[4]A theoretical use of the terms state and nation would have lent Kaldor's research the depth that it currently lacks. Neither in Yugoslavia nor during the nation-building processes of the 1700s and 1800s can the term state be said to have been more of a driving force for violence than the term nation.

[5]van Creveld (1991) and Keegan (1993).

[6]Foreign enclaves deep in the realm disappeared. An example of this is Avignon in the centre of France, many German free cities disappeared, in fact, only four free cities remained of the countless numbers that existed before the Napoleonic Wars.

[7]In addition, it must be stated that Kaldor's closing chapter, which presents a vision of a cosmopolitan world order, is at the expense of historical experience. When Kaldor encourages the reader to incline towards outstanding researchers such as Zygmunt Bauman and Norbert Elias on one side or the author herself on the other, then the choice easily falls in favour of the overwhelming force of evidence presented by the former. The two former are not pessimistic because it pleases them. Their pessimism is founded on a sound knowledge of mankind and its history.

Demonstrating theory using empirical examples is not the same as an empirical application of the theory. It is worth differentiating here; if the lead up to the conflict in former Yugoslavia had been understood in field theory terms at both the strategic and tactical level then the scenario may well have had a different character—it is even possible that the course of events would have unfolded in a different way. A related approach, though not the same, to the events has been taken by the researcher V.P. Gagnon. He has picked up a somewhat poststructuralist view when he writes that:

> Elites who are highly dependent on the existing structures of power, and for whom change would mean a total loss of access to and control over resources, will be much more willing to pursue strategies that are extremely destructive to society overall.[8]

This holds some truth and certainly in the example of Yugoslavia. But society must present a social space, which allows those strategies to be used in a legitimate way. Furthermore, the actors must be structured so that they actually take the opportunity to use this space of possibilities which the structure of society present. In short, both the actors and society need a certain structure, otherwise, the violent scenario does not play out. This is the closest to a law we can come. If it was not for the play between society and actors then such a case as the one in Yugoslavia would play out more often than it does. Michael Mann is also preoccupied with the question of responsibility:

> If a few bad guys were responsible, how did they acquire such magical powers of coercion and manipulation? And were they quite so coherent in their planning; so in charge of events? After all, atrocities were committed by thousands of persons, and many more thousands stood around, either egging them on or doing nothing to stop them.[9]

On the other hand, he does not—compared to Gagnon but certainly not compared to this text—venture very much into the structuralism

---

[8] Gagnon (2004, p. 29).
[9] Mann (2005, p. 360).

explanations. Nevertheless, the question is valid and fits well into the perspective of social field theory. Nationalism in various forms was not an unknown phenomenon in Yugoslavia, which came into being after the First World War.[10] Often nationalism is placed in a political right corner, in opposition to, for example, communism. However, nationalism does not hold any values, which make it a given rightist entity. Nationalism is about power and can be a tool in anyone's hand as long as legitimacy is granted. Nationalism was present in communist countries, even the official communistic architecture contained nationalistic features.[11] In both countries, the issue of nationalism was complex, both having a population consisting of many different national identities. In Yugoslavia, the situation was further complicated by the fact that the official war heroes, the partisans revered from recent history, were intimately associated with the nation. Furthermore and more importantly, they were associated with communist ideology since the partisan movement—the People's Liberation Movement (*Narodnooslobodilački pokret, NOP*)—was after all Marxist. When communist ideology lost its legitimacy after 1989, the prestige (symbolic capital) enjoyed by the partisan movement also disappeared. The state of Yugoslavia had to a large extent built its legitimacy on this now-eroded ideological foundation. Yugoslavian nationalism was not very strong, which can easily be explained by the fact that it was a state with many nations, i.e. not a national state but a state with nations. In Yugoslavia, alternatives lay just below the surface, especially in the form of the earlier defeated movements in Serbia and Croatia. Yugoslavian nationalism had problems finding any support as there were alternatives with stronger roots. An outstanding opportunity to fill the political vacuum with new forms of capital therefore presented itself for those prepared to take it after the fall of the Eastern Bloc: powerful nationalist forces fuelled by crisis and approaching anarchy boosted by the fall of communism. This structural change of the world political order was important for the way Yugoslavia was structured, as for all of Eastern Europe.

---

[10] First called "Kingdom of Serbs, Croats and Slovenes".

[11] Kaldor (1999, p. 79). Mary Kaldor, like many others, has emphasised the importance of nationalism both in the Soviet Union and in Yugoslavia.

Both the Serbian Chetnik and the Croatian Ustasha movements had been suppressed by the conquering communist power. There was therefore a need to reinvent the cultural area, which was called Yugoslavia. It was a process about getting rid of the effects of decade long symbolic violence against the historic opposition to the fallen communist regime to which those who sympathised with Serbian and Croatian nationalist ideals could affiliate themselves with. In turn, the economic crisis that followed the collapse of the communist system also provided a fertile breeding ground for extremist ideas. As a result of the nationalist movements of their recent past, both countries were strong, and the Serbs' and Croats' unwillingness to share power in Yugoslavia, in a way that was acceptable to both, served to weaken the cohesion of the rather weak Yugoslav nationalism.[12] The period between the early 1980s with Tito's death and the early 1990s can be described as a period of ethnification in Yugoslavia, where, for example, promotion of Serbian symbols in Serbia became stronger over time. One can talk about a mobilisation of history for political purposes, a construction of the past.

It is important to point out that these nationalistic expressions could not be traced back over hundreds of years; they were not of ancient origin. It did not have its origins in conflicts inspired by cultural determinism. The reinvented opposition against Yugoslavia was founded in the Yugoslavia of the 1900s, which had seen so much criminality.[13] The ethnic cleansing of the type that occurred during the 1900s in Yugoslavia and in Bosnia in particular was not a necessary step in the national deconstruction and reconstruction processes, even in the Balkans. There were underlying factors, such as the nationalist movements mentioned above and the crimes committed during the Second World War, which bedded for the events of the 1900s.[14] These factors described the logic

---

[12]For dissent between Serbs and Croats see also Naimark (2001, p. 140).

[13]Cohen (1993, p. 238). It is also emphasised here that hatred between ethnic groups did not exist before the Second World War. The Nazi Ustasha state should be mentioned as a major actor, as the Chetnik movement and the partisans but also the Albanian oppression of Serbs in the Italian controlled Kosovo.

[14]As a reference for someone who saw the crimes committed against the Bosnian Muslims as a religious issue: Pasha Mohamed Ali Taeharah. *An Introduction to Islamism.* Author House (2005, p. 24).

of practice, which the reinvented movements could draw legitimacy from. The logic of practice was no more precise than that ethnic cleansing was a viable alternative; it could then be expanded to new conflicts, as the one between Bosnians and Croats.

One can at this point question the relevance of stressing that opposition did not stretch back over hundreds of years. Certainly, there is a point in putting an end to the myth that a desire to wage civil war is something that lies in the genes of any people—it is an important factor, but if one is interested in the events of the 1900s, one cannot minimise the fact that the Second World War is part of the collective *habitus* even for those who didn't experience it. Of course, the exact implications of this depend on what kind of narrative the actors (collective or individual) are being fed. At the same time, the model supports a structural perspective that the actors are given the freedom to choose their own approach, regardless of the effect of the structure. It is likely that Yugoslavia would have fallen apart regardless of who held political power. However, without the conscious policies of Milošević and Tudjman, extreme violence would not have marked the course of events.[15] Discussing which one of them was foremost or more of a driving force is in this context immaterial, since it is not a question of apportioning guilt, but rather than stressing the dynamic inherent in the fact that at least two different political themes—one about centralise Yugoslavia under a single command and one about making Croatia autonomous—could paradoxically support each other.

However, in Milošević's case, one can say that his policy of radicalisation was the foundation of the power base, he later established in Serbian politics. From having held a strong, but not unique, position with communist leanings in Parliament, he became the leading Serbian politician with a strongly nationalist manifesto.[16] From Milošević's rise to power in 1987, the media in Belgrade was very much in his hand, at

---

[15] Naimark (2001, p. 139). See also Donia and Fine (1994, p. 11). The claim is made that the Second World War was the first occasion when ethnic cleansing occurred in Bosnia.

[16] This is described more fully by Sell (2002). Particularly Chap. 2 for Milosevic's radicalisation programme and rise to power and Chap. 3 for when the agenda takes a more violent form.

least up to 1990. Milošević was among the most prominent leaders of the Communist Party during the late 1980s, but he also saw the perceived rift between Serbian interests and communism. Some observers have thought that "communism was viewed as the thing that was weakening the Serbs' position".[17] Another observer has written that Milošević "seemed to stumble almost by accident on the nationalistic card in April 1987, in the small town of Kosovo Field, next to the famous battlefield".[18] He followed up with his speech at Gazimestan—the battlefield 600 years ago of mythic proportions in Serbian history. "Six centuries later we are once again in battles, and facing battles. They are not armed battles, though the possibility cannot be excluded".[19] Milošević had clearly adopted a more nationalistic approach, which was in line with the currents of politics in all of Eastern Europe at the time. He had already been one of the actors able to decide the agenda, but now as the *producer* for national politics, he listened attentively to the views prevalent in the strong *consumer* currents of Serbian politics at the time, which (greatly) enhanced his position. To be fair towards Milošević, one should mention that although he had consecrating power on at least the Serbian political field, there were also other political actors who encouraged extremism. The poet and the then not really active in politics Vuk Drašković, for example, raised the provocative question in 1989 "where are the Western borders of Serbia, and how far do they extend?"[20] In this case, there is structuring structures at work with a political field steering actors into certain behaviour who in their turn structure the field—mutually strengthen each other. In addition, Milošević was more important than Tudjman, because Serbia was to a marked degree the cement holding Yugoslavia together—not least because the capital city

---

[17] Pavlakovic´ (2005, pp. 2, 16).

[18] Mann (2005, p. 369). I do not mean that the Balkans have a stronger willingness to embrace myths than for example Western Europe. I state that there were myths in play—and probably still are—and that is the state of most cultures. See (Todorova 2005, p. 153).

[19] Quoted in Mann (2005, p. 370).

[20] Stojanovic (2000, p. 462).

lay in Serbia. If Serbia did not recognise the union, then there would be little incentive for it to hold together, so what happened in Serbia then mattered very much more than what happened in Croatia. These two politicians have been described as Tudjman being the more "fanatical nationalist" and Milošević as the opportunistic one.[21] It is also true that Tudjman had a stronger position in Croatia than Milošević had in Serbia. A position built on Tudjman and this party HDZ: "dictating a political discourse of authoritarianism and xenophobic nationalism".[22] Tudjman had a background of radical nationalism, serving two terms in jail due to Croatian nationalistic activity.[23] Robert Hayden discusses Tudjman's view on nations and nationalism in the terms that the view could easily be flipped into being a racist view of looking at nations.[24] This in its turn gives a greater understanding of how the coming ethnic cleansings in Croatia came about.

Presented with a specific empirical situation, the actors are given a horizon of possible courses of action to follow. In a radical situation such as a state of war, individuals will react within the framework in a way unique to that situation. This may mean that many in a civilised society react by, for example, fleeing to another country. But the underlying and radical cultural manifestation, which makes up part of an individual's habitus, may lead to individuals either collectively or individually reacting violently; despite the fact that a short while ago to all appearances they could not be told apart from anybody else on the street. It concerns cultural dispositions, which are not always that easy to identify except in retrospect. The approach both differs from and has interplay with the concept of external shock. Researchers into military culture often subscribe to the widely held opinion that external shock is necessary to undermine the legitimacy of cultural norms.[25] It is therefore also likely that a trigger factor will be required if a change to cultural norms is to be effected.

---

[21] Allin (2002, p. 22).
[22] Sekulic et al. (2006, p. 808).
[23] Sell (2002, p. 115).
[24] Hayden (1992, p. 663).
[25] Farrell (2005, p. 14).

However, one cannot be at all sure that structures within a culture (not the same as cultural norms) do not exist, which may well incline collective or particular individuals to violent behaviour. It should also be borne in mind that any intense experience will contribute to the forming of an individual's habitus. Violent events thus serve to stretch *the bounds of what for an individual is capable of* so that previously inaccessible violent agendas are afforded space. Christopher Browning's account of the action of the Reserve Police Battalion 101 in Poland is an example of this.

At this point, a résumé of field theory is useful. This states that those who really possess the power to change the *illusio* of the field, the actual definition of the field, are the actors with strong capital on the left-hand side of the field.[26] The left side of the field is the *autonomous* area, where the actors/institutions that play according to the field's own rule are to be found. On the right are the actors who follow rules other than those prescribed by the field. In the political field constituted by Yugoslavia, Milošević and Tudjman were by far and away the strongest political actors. Initially, they operated in the upper left quadrant of the field. By virtue of their prestige, they had the power to change the rules of the field. The nationalist agenda of both individuals lent legitimacy to their conduct of politics by violent, rather than peaceful means. The pursuance of policies using violent means was an unknown strategy for the left side of the field, at least until the most respected actors in the upper part of the left field began lending an air of legitimacy to violent tactics. One can compare with a statement regarding Iraq: "in the general's words, the ethnosectarian violence of 2006 had torn apart the very fabric of Iraqi society".[27]

A further consequence of Tudjman's and Milošević's actions was that if the most powerful actors on the left side of the field weakened the appeal of politics by constitutional means and further put force behind a violent approach, then this would affect the regard in which the remaining political actors were held. The actors who continued to choose the constitutional path found their prestige increasingly eroded, while those who altered their political approach in line with the new violent laws of the field found themselves rewarded—to say nothing of the actors who occupied a permanent position on the right side of the field.

---

[26]Consecrating power, the laying on of hands, which certain powerful actors possess as a result of their prestige (symbolic capital) and/or their official position. These actors have the power to determine the value of other actors or positions on the field.

[27]Kilcullen (2009, p. 131).

It should be noted that the friction between the ethnic and the constitutional models was already discernible in Serbia at the time of the First World War, there being few with Yugoslavian inclinations among the Serbian intellectual elite.[28] This should be borne in mind, since Milošević's main political opponents in Serbia were also extreme Serbian nationalists like, for example, Vuk Drašković and Vojislav Šešelj.[29] The latter built the Chetnik movement into a fighting force when the violence began to take hold.[30] This may serve as a clear example of how a politician switches from a civil political agenda to one of violence, as a result of the political rules of the field changing. As one can see, there were underlying spaces of possibilities which opened up with the fall of the Eastern Bloc. Much of the political agenda reinvented a political field, which drew legitimacy from the field as it looked like before the communist era.

BIBLIOGRAPHY

Allin, D. H. (2002). *Nato's Balkan interventions*. Oxford: Routledge.
Behschnitt, W. D. (1980). *Nationalismus bei Serben und Kroaten 1830–1914. Analyse und Typologie der nationalen Ideologie*. Oldenburg: Oldenbourg Wissenschaftsverlag. https://www.cia.gov/library/publications/the-world-factbook/geos/bk.html [Visited 170509].
CIA Factbook. (2017). https://www.cia.gov/library/publications/the-world-factbook/ [Visited 170509].
Cohen, L. J. (1993). *Broken bonds. The disintegration of Yugoslavia*. Boulder: Westview Press.
van Creveld, M. (1991). *The Transformation of War*. New York: The Free Press.
Donia, R., & Fine, J. (1994). *Bosnia and Herzegovina. A tradition betrayed*. London: Colombia University Press.
Farrel, T. (2005). *The norms of war. Cultural beliefs and modern conflict*. Boulder: Lynne Rienner.
Gagnon, V. P. (2004). *The myth of the ethnic war. Serbia and Croatia in the 1990s*. Ithaca: Cornell University Press.
Hayden, R. (1992). Nationalism in the former Yugoslav Republics. *Slavic Review, 51*(4), 654–673 (Cambridge).
Kaldor, M. (1999). *New and old wars*. Cambridge: Cambridge University Press.
Keegan, J. (1993). *A history of warfare*. New York: Vintage Books.

---

[28] Behschnitt (1980, p. 233). See also Miller (1997, p. 179).

[29] Naimark (2001, p. 153).

[30] Sikavica (1997, p. 141).

Kilcullen, D. (2009). *The accidental guerrilla. Fighting small wars in the midst of a big one*. Oxford: Oxford University Press.

Krampe, F. (2010). Neue Kriege, neu betrachtet. Neubetrachtung des Forschungsstands und des Fallbeispiels Bosnien und Herzegovina. *Zeitschrift für Genozidforschung, 10*(1): 61–92 (Bochum: Ruhr Universität).

Mann, M. (2005). *The dark side of democracy. Explaining ethnic cleansing*. Cambridge: Cambridge University Press.

Miller, N. J. (1997). *Between nation and state. Serbian politics in Croatia before the First World War*. Pittsburgh: University of Pittsburgh Press.

Naimark, N. M. (2001). *Fires of hatred. Ethnic cleansing in twentieth-century Europe*. Boston: Presidents and Fellows of Harvard College.

Pavlakovic, V. (2005). Serbia transformed? Political dynamics in the Milošević Era and after. In S. P. Ramet & V. Pavlakovic (Eds.), *Serbia since 1989. Politics and society under Milošević and after*. Seattle: University of Washington Press.

Sekulic, D. et al. (2006). Ethnic intolerance and ethnic conflict in the dissolution of Yugoslavia. *Ethnic and Racial Studies, 29*(5), 797–827 (New York: Routledge).

Sell, L. (2002). *Slobodan Milošević and the destruction of Yugoslavia*. Durham: Duke University Press.

Sikavica, S. (1997). The army's collapse. In J. Udovicki & J. Ridgeway (Eds.), *Burn this house. The making and unmaking of Yugoslavia*. Durham: Duke University Press.

Stojanovic, D. (2000). The traumatic circle of the Serbian opposition. In N. Popov (Ed.), *The road to war in Serbia. Trauma and catharsis*. Budapest: Central European University Press.

Todorova, M. (2005). The trap of backwardness: Modernity, temporality and the study of Eastern European nationalism. *Slavic Review, 64*, 140–164.

CHAPTER 6

The Events in 1990

Abstract This chapter deals with the continuation of the conflict and the breaking up of Yugoslavia. By 1990, Federal Yugoslavia was unpopular. Most Yugoslavs wanted to move from communism to democracy, yet they associated federation with communism and Serb domination.

Keywords Yugoslavia · 1990 · Communism · Serbia

Outside of Serbia, almost all wanted decentralising reforms, but most Serbs disliked the decentralisation that had already occurred.[1] The League of Communists of Yugoslavia collapsed on 20–22 January 1990, when Slovenia and Croatia left the Congress that was held. When SFRY (Social Federal Republic of Yugoslavia) collapsed, SPS (Socialist Party of Serbia) stressed that where Serbs were in majority, they should be able to say that they wanted to remain in the Yugoslavian state. In March 1990, a new Serbian constitution was ratified, limiting the autonomy

---

[1] Mann (2005, p. 366).

© The Author(s) 2017
H. Gunneriusson, *Bordieuan Field Theory as an Instrument for Military Operational Analysis*, New Security Challenges, DOI 10.1007/978-3-319-65352-5_6

of Vojvodina and Kosovo.[2] This was a strike against Yugoslavia as an entity. No new president could be elected for Yugoslavia in the early 1990s because Milošević controlled the republics and blocked the sole presidential candidate all according to the parliamentary system of Yugoslavia. The position Bosnia, Croatia, Macedonia and Slovenia voting for the presidential candidate, and Serbia, Vojvodina, Kosovo and Montenegro voting against. [3] This process has been described as a slide from centralism towards confederalism, which is true. This change of the political field also had an impact on the military. It resulted in the JNA (Jugoslovenska Narodna Armija, i.e. The Army) partly redefining its role and eventually its conception of itself.[4]

It is interesting to note that the republics that were soon to break away, Slovenia and Croatia, were at this stage still in favour of holding Yugoslavia together, even if it looked as if later they would separate from the federation. If that presidential election had had a positive result, with the installation of a president, there would still have been accepted constitutional structures in place to lean back on during the partitioning process, which would have led to a more peaceful course of events. As it turned out, relatively undeveloped areas like Kosovo (and even Montenegro might be regarded as undeveloped) came to stifle the desire to move in that direction in more developed areas like Croatia and Slovenia—and in comparison especially with Kosovo, in the more industrialised Bosnia. This created what is usually known as a democratic deficit, albeit the term is more appropriately used in connection with established democracies. Nevertheless, the situation generated a feeling of inherent injustice and illegitimacy towards the constitution of Yugoslavia. This ought to have contributed to strengthening the desire of Croatia and Slovenia to go their own way. Meanwhile in Bosnia—after the elections of 1990—extremism came in focus in Bosnia. For example, Karadic's Serbian Democratic Party (SDS; Srpska Demokratska Stranka)

---

[2] Kerenji (2005, p. 367).

[3] Donia and Fine (1994, p. 214). Dyker and Vejvoda (1996), claim that Croatia for the most part blocked all possibility of a federal election. p. 19.

[4] Dulić and Kostic (2010, p. 1064).

slid from a moderate line towards a more extreme one.[5] After the weak results in the election, SDS became more nationalistic speaking of the western borders of Serbia and saying that they wanted peace, but the current status was more like capitulation than peace.[6]

The ones who had actively supported a united Yugoslavia were marginalised by the obvious hopelessness of trying to elect a federal president, and the surge of confederalistic or even separatist rhetoric. Expressed in theoretical terms, this marginalisation can be viewed as a loss of capital, the ability to influence the political field being reduced, while those with agendas other than a united Yugoslavia gained wider room for manoeuvre. Milošević and SPS presented themselves as being a more moderate alternative to many other parties from 1990 onwards.[7] By the SPS victory with 45.8% on 9 December 1990, the membership had risen largely due to new members. The SPS went to election with slogans of peace and prosperity, in opposition to more nationalistic alternatives. Serb communists moved towards nationalism in order to prevent giving the opposition monopoly of it. The restructured field forced restructuring on the actors, and the space of possibilities for them changed accordingly.

Kosovo was the start for an expression of opinion along Serbian lines which then surfaced in other parts of Yugoslavia. The events in Kosovo in 1990 signalled to the whole of Yugoslavia, and certainly to Slovenia and Croatia, that Milošević not only practiced nationalist rhetoric but also had a coordinated policy line. On 8 of September, the Serbian constitution was changed, drastically reducing the autonomy previously enjoyed by Vojvodina and Kosovo.[8] This upsets the legitimacy of the

---

[5] Gagnon (2004, p. 50).

[6] Stojanovic (2000, p. 469). May 1991 was the time when Mirko Petrovic talked about the western borders.

[7] Gagnon (2004, p. 46). In the 1992 election, Milosevic was challenged by Milan Panic in selling the moderate line. Ibid, Gagnon noting a working paper of his.

[8] Udovicki and Torov (1997, p. 92).

political system and thus made way for alternate ways of carrying out politics. It should be noted in this connection that there were a number of background factors that made Milošević's nationalist agenda viable, but it was his political strategy that was the initiating and driving force. A similar question is whether a German war of revenge would have taken place had it not been for Hitler. The answer is probably given, the harsh Versailles treaty. Would the Holocaust have taken place without Hitler? The answer is no, at least no if the Nazis did not come to power but some other right-wing movement with no specific anti-semitic agenda.[9] The comparison is made not to put Milošević on a level with Hitler, which would be outrageous, but to show that structural change often is difficult, regardless of which actors are involved. Both the Second World War and the partitioning of Yugoslavia were examples of this phenomenon. On the other hand, the ethnic cleansing in Yugoslavia and the Holocaust during World War II would be hard to imagine without actors that at least had the same type of agenda as the two leaders named above.

Both the Serbian and Bosnian nationalist movements repeatedly referred to the Second World War, in order to create a sense of continuity for and lend legitimacy to their own movements. This approach is relatively common, regardless of whether it concerns politics, science, business or other areas. Institutions or people who had high reputations in the past in the eyes of a particular group will still be of current value and also used because of that very value. It is a question of identifying oneself with the symbolic capital of the actor or institution concerned and thereby strengthening one's own position at the same time.[10] Tudjman was an accomplished politician of the Realpolitik genre and may have seen that these Nazi references were favourably regarded by the Croatian diaspora.[11] Many of these had fled Yugoslavia after the Second World War, and even if new generations had come, the nationalistic master narrative sprung from the Ustasha Croatia was strong in the diaspora. The same can be said about the Serbs to some extent. Serbia started to run a satellite TV channel of its own, mostly because

---

[9] This argument is first made by the nazi-German dissident Sebastian Haffner. Haffner (1991, p. 216).

[10] Gunneriusson (2002, p. 38).

[11] Dulić (2009, p. 263).

they wanted to reach the Serbian diaspora. Many of those had emigrated because of rightist Chetnik sympathies in the generation before or even the same generation.[12] They were disposed to be affected by the propaganda. Both diasporas had power not only because of their ability to work on the opinion abroad which is a form of PSYOPs by proxy, but also that they had more financial assets than those in the former Yugoslavia had at the time.

It was of no great concern that these movements might not gain legitimacy in the eyes of the international community in general since their policies were geared for those who identified themselves as Croatian in general. This almost paradoxical form of connection occurred at several levels. Franjo Tudjman (created Croatian President on 30 May 1990) declared himself ready to cast off the Nazi yoke that had lain over the Croatian state since the Second World War, when Croatia had been a satellite state. According to Tudjman, this would best be achieved by destroying the memorials to the victims of the outrages committed in the earlier Croatian state.[13] The process involved partly removing all trace of the exponents of the former system's symbolic capital, and partly served to increase the capital of Franjo Tudjman himself. Paradoxically enough, making a connection between the former and current states of Croatia was unnecessary, since the budding Croatian state was a new state; an opportunity for a fresh start was lost. Tudjman made television in Croatia a part of his party's (HDZ) domain as soon as he came to power. The press was also brought under control relatively effectively.[14] The information arena was not neglected as a means to hold and reinforce power.

Tudjman was the one who made the equation between the two Croatian states. In contrast to casting off the Nazi yoke, he tied the old state's identity to that of the new state. The second string to Tudjman's ultranationalist bow was to manipulate the historiography of genocide of the Serbian people committed by the Ustasha state during the Second World War. The researcher Tomislav Dulić writes that Tudjman used three different arguments to downplay the atrocities committed by Ustasha. Firstly, Tudjman states that there have been Serbian

---

[12] Dimitrijevic (2000, p. 638 and notes 17 and 18).

[13] Udovicki and Torov (1997, p. 111).

[14] Balas (1997, p. 266).

exaggerations; secondly, that some sources have been at fault; thirdly, that he relativises the events.[15] How and to whom he catered these theses is one thing, but it does also say something about how Tudjman was structured and how his space of possibilities was constructed. Since the 1950s, he had worked on a revisionist history to reduce the blame put on the Ustasha Croatian state during World War Two.[16] This was not a new idea for him, based on pure opportunism. Apparently, he did not see the actions of the Ustasha as very wrong so one can assume that he was disposed to not only forgive such actions but also consecrate them if the right situation appeared again for something similar to happen.

This resulted in morbid, but certainly, necessary countermeasures being taken by the Serbs when Serbian mass graves were dug up in the summers of 1989 and 1990 in Krajina, Croatia, to counteract Tudjman's falsification of history.[17] The Krajina area in Croatia was used as a token which both politicians played on, as Tudjman wanted Croatian autonomy, and Milošević said that it was impossible with the old Serbian settlement Krajina within its borders.[18] The region was an old Serbian settlement, called "the military border" where the Croatian Ustashi regime conducted genocide against Serbs during World War Two.[19] After the elections in 1990, the Serb leadership in the region proclaimed the area an autonomous region (an *oblast*) which came to have different names during its existence.

Tudjman also denied Croatia's part in the Holocaust during the Second World War.[20] In addition, Franjo Tudjman declared that the Ustasha state was a worthy predecessor to modern Croatia.[21] This type of unwholesome retrospective historical connection enabled these former

---

[15] Dulić (2009, p. 264).

[16] Dulić (2009, p. 278).

[17] Loc. cit. This was mainly a success as the US policy included the ethnic cleansing of Krajina, where Serbs had lived for 500 years. The Americans saw it as "recapturing the territory from the Serbs". Allin (2002, p. 30).

[18] Glenny (1992, p. 37).

[19] Sell (2002, p. 113).

[20] Ramet (1999, p. 51).

[21] Naimark (2001, p. 154).

criminal deeds to move from the era of the Second World War to the surface of the contemporary political agenda. Croatian units also used Ustasha insignia, which strengthened ties with the past and contributed to eradicating the difference between the present and the past.[22] HDZ success came from playing on threats from Belgrade and on the surge for alternatives to communism.[23] In addition, Tudjman's policies lent legitimacy to Milošević's policies, by confirming that which aggressive nationalism was saying the Serbs had been subjected to. This led to an undermining of the reputations of those who opposed Milošević in Serbia, when he was seen to be obviously right, given the context, in the light of Tudjman's actions.

Budding Serb nationalism was, paradoxically enough, the very breath of life for the politics of Tudjman in Croatia and vice versa.[24] Serb nationalistic intellectuals provided further arguments along the lines of ethnicity and nation.[25] Croatia had the support of the West from the very start and right through to the end of the conflict, in contrast to the rest of Yugoslavia. This is noteworthy bearing in mind what the country stood for and did; the regime was virtually a mirror image of the more vocal Serbian nationalistic politicians. These are the central elements to the understanding of the internal conditions in Yugoslavia before violence came to the surface. The same type of mutual relationship was visible between Likud and the PLO, especially under the leadership of Arafat. The latter had seen his reputation weakened among the Palestinian people, but he enjoyed an upswing during the unrest of 1996. Arafat as well as Likud needed an external enemy in order to be able to use the image in their domestic arenas.[26] So, one can see that there was a drive towards radicalisation in politics before the major hostilities broke out.

In Serbian politics, Milošević could take steps towards the break-up of Yugoslavia and eventually war, much because he was not very extreme

---

[22] Naimark (2001, p. 157 (about Second World War references being used) and p. 172).

[23] Gagnon (2004, p. 47).

[24] For the interlinking connection between the politics of Tudjmans and Milosevic, see, for example, Udovicki and Torov (1997, p. 93) and Stitkovac (1997, pp. 156 and 158), also Udovicki and Stitkovac (1997, p. 174).

[25] Sell (2002, p. 111).

[26] Hammes (2006, p. 117).

in comparison. Vojislav Šešelj and Vuk Drašković went on with different radical projects. On 18 June 1990, Vojislav Šešelj founded the Serbian Chetnik Movement as an attempt at a party, but authorities refused to register it as a party.[27] Vuk Drašković became the leader of SPO (The Serbian Renewal Movement, *Srpski Pokret Obnove*) which he started. In the beginning, SPO was formed as a pyramid organisation with an "unimpeachable leader".[28] Before the war in Croatia, in May 1990, the SPO wanted autonomous Serbian regions in Krajina, Istria, Dubrovnik and 4 regions in Bosnia.[29] On 7 January 1990, Vuc Drašković proclaimed that the goal of his party was "the creation of a democratic, multiparty Serbian state within her historical and ethnic borders".[30] From July to December 1990, one-third of SPO's statements were about the national question. Correspondingly, only 6% of SPS and DS statements concerned this.[31]

The role of the Orthodox Church could be mentioned as the text deals mostly with the politics in Serbia, and the church indeed played a political role on the social field of politics in Yugoslavia. Bishop Simeon Zlokovic´ was a critic of both Tudjman and Milošević. He saw them both as representatives of extremes on the right–left political scale. This was in June 1990, and Milošević was not then very much a traditional communist, but the bishop did rightly see these two actors as the symbols for extreme politics, and in hindsight, he was right. As the Orthodox Church and the myths and history of Serbia are intertwined, it is easy to see that the church had a nationalistic profile. The Orthodox Church in Serbia is by its nature tightly linked to Serbian nationalism, and vice versa.[32] History is what the Orthodox Church and Serbian nationalism had in common. The Church did ask for Serbian unity in the elections of 1990 and warned against genocide of Serbs and Ustasha

---

[27] Thomas (1999, p. ix).

[28] Stojanovic (2000, p. 455).

[29] Stojanovic (2000, p. 462).

[30] Stojanovic (2000, p. 463).

[31] Stojanovic (2000, p. 468).

[32] Ramet (2005, p. 256).

in Bosnia.[33] The connection between culture and land was strong in the church's rhetoric; for example, the church asked the Bosnian Serbs to stay in "their ancestral homes".[34] The church also came to be in a state of denial further on, as it denied the existence of concentration camps run by Serbs in Bosnia.[35]

So, the Orthodox Church was very much linked to nationalism in Serbia. Belief certainly played a part in Croatia too. The change in the figures among declared believers in Croatia rose from 47% in 1989 to 76% in 1996.[36] In the case of the Muslim population of Croatia, one just has to mention that much of the violence imposed on these Muslims came into force just because they were Muslims. This of course moulded them together and strengthened the importance of being a Muslim in Bosnia and not just being a Bosnian. Nothing of this suggests that religion encourages ethnic cleansing. But religion was important in creating a *we* and by that also creating a *they*. This in its turn had consequences when it came to structuring Yugoslavia into a violent place during the 1990s.

During 1990, Yugoslavia as a project appeared to be a lost cause, and the alternatives grew in strength in both Serbia and Croatia. The nationalistic rhetoric increased in Croatia, but in Serbia, smaller parties followed the same road, opening up for Milošević to apply a more nationalistic approach without looking all too extreme—the latter would have scared popular support away. Worth mentioning is that it was not only the end of the Cold War that was paramount for the change taking place but the lack of a democratic heritage in the new multiparty state also played a role. Even if Yugoslavia under Tito had been a rather benevolent totalitarian state, it still was a totalitarian state which structured its population and politicians. All in all, the political field was restructured by actors whose space of possibility in the given situation had changed.

---

[33] Ramet (2005, p. 258). See also p. 262 about church resistance against Milosevic.

[34] Ramet (2005, p. 259).

[35] Ramet (2005, p. 258).

[36] Sekulic et al. (2006, p. 814). See also p. 818 about religion being important in both Serbian and Croatian nationalist ideology.

BIBLIOGRAPHY

Allin, D. H. (2002). *Nato's Balkan interventions.* Routledge: Oxford.

Balas, S. (1997). "The opposition in Croatia". Burn this house. The making and unmaking of Yugoslavia. In J. Udovicki & J. Ridgeway (Eds.). Durham: Duke University Press.

Dimitrijevic, V. (2000). Yugoslavia and the world. In N. Popov (Ed.), *The road to war in Serbia: Trauma and catharsis.* Budapest: Central European University Press.

Donia, R., & Fine, J. (1994). *Bosnia and Herzegovina. A tradition betrayed.* London: Colombia University Press.

Dulić, T. (2009). Mapping out the wasteland: Testimonies from the Serbian commissariat for refugees in the service of Tudjman's revisionism. *Holocaust and Genocide Studies, 23*(2): 241–246. (Oxford).

Dulić, T., & Kostic, R. (2010). Yugoslavs in arms: Guerilla tradition, total defence and the ethnic security dilemma. *Europe-Asia Studies, 62*(7): 1051–1072. (New York: Routledge).

Dyker, D. A., & Vejvoda, I. (1996). *Yugoslavia and after. A study in fragmentation, despair and rebirth.* New York: Routledge.

Gagnon, V. P. (2004). *The myth of the ethnic war. Serbia and Croatia in the 1990s.* Ithaca: Cornell University Press.

Glenny, M. (1992). *The fall of Yugoslavia.* London: Penguin Books.

Gunneriusson, H. (2002). *Det historiska fältet. Svensk historievetenskap från 1920-tal till 1957.* Uppsala: Studia Historica Upsaliensia.

Haffner, S. (1991). *The ailing empire: Germany from bismarck to Hitler.* New York: Fromm International.

Hammes, T. X. (2006). *The sling and the stone: On war in the 21st century.* St. Paul: Zenith Press.

Kerenji, E. (2005). Vojvodina since 1988. In S. P. Ramet & V. Pavlakovic (Eds.), *Serbia since 1989. Politics and society under Milošević and after.* Seattle: University of Washington Press.

Mann, M. (2005). *The dark side of democracy. Explaining ethnic cleansing.* Cambridge: Cambridge University Press.

Naimark, N. M. (2001). *Fires of hatred. Ethnic cleansing in twentieth-century Europe.* Boston: Presidents and Fellows of Harvard College.

Pavlakovic, V. (2005). Serbia transformed? Political dynamics in the Milošević Era and after. In S. P. Ramet & V. Pavlakovic (Eds.), *Serbia since 1989. Politics and society under Milošević and after.* Seattle: University of Washington Press.

Ramet, S. P. (1999). *Balkan Babel. The disintegration of Yugoslavia from the death of tito to the war for Kosovo.* Boulder: Westview Press.

Ramet, S. P. (2005). The Politics of the Serbian Orthodox Church. In Sabrina P. Ramet & Vjeran Pavlakovic (Eds.). Seattle and London: University of Washington Press.

Sekulic, D., et al. (2006). Ethnic intolerance and ethnic conflict in the dissolution of Yugoslavia. *Ethnic and Racial Studies, 29*(5): 797–827. (New York: Routledge).

Sell, L. (2002). *Slobodan Milošević and the destruction of Yugoslavia.* Durham: Duke University Press.

Stitkovac, E. (1997). Croatia. The first war. In J. Udovicki & J. Ridgeway (Eds.), *Burn this house.* The making and unmaking of Yugoslavia. Durham: Duke University Press.

Stojanovic, D. (2000). The traumatic circle of the Serbian opposition. In N. Popov (Ed.), *The road to war in Serbia. Trauma and catharsis.* Budapest: Central European University Press.

Thomas, R. (1999). *Serbia under Milošević politics in the 1990s.* London: C Hurst & Co Publishers Ltd.

Udovicki, J., & Stitkovac, E. (1997). Bosnia and Hercegovina: The second war burn this house. In J. Udovicki & J. Ridgeway (Eds.), *The making and unmaking of Yugoslavia.* Durham: Duke University Press.

Udovicki, J. & Torov, I. (1997). "The interlude". Burn this house. The making and unmaking of Yugoslavia. In J. Udovicki & J. Ridgeway (Eds.). Durham: Duke University Press.

CHAPTER 7

The Events in 1991

Abstract This chapter deals with the final blows against Yugoslavia as the state it once was and its eventual reduction to an extension of Serbian politics. From the point of legitimacy, one can see this as the height of Milošević political career, even if his power would be strong well beyond this year. Serbian forces attack eastern Croatia and sack the city of Vukovar as the first major hostilities between the two parts of the former Yugoslavia.

Keywords Milošević · Serbia · Croatia · Vukovar · Yugoslavia

Borisav Jovic stepped down as president of SFRY on 15 March 1991, when martial laws were turned down—something he advocated given the situation. By the same time, March 1991, Milošević had abandoned federalism (Plan A), instead of seeking to enlarge the Serbian-controlled territory (Plan B). He repeatedly called for "All Serbs in one state". The code for Plan B was the military line, *Vojna Linija,* which meant covertly arming the Serb *precani* communities, more about that to follow.[1] On 25 March 1991, Milošević and Tudjman met in secret in Karadjordjevo and made up common plans for dividing BiH (Bosina and Herzegovina),

---

[1] Mann (2005, p. 390). Precani basically means western Serbian settlements, e.g. in Croatia.

H. Gunneriusson, *Bordieuan Field Theory as an Instrument for Military Operational Analysis,* New Security Challenges, DOI 10.1007/978-3-319-65352-5_7

at the expense of Muslims, but reached no agreement.[2] Tudjman did not initially favour Croatian independence for pragmatic reasons: a former army general, he feared a JNA invasion. So while bargaining, he was covertly seeking arms and military advisers abroad (as Izetbegović in Bosnia did not). The longer the delay, the more he could arm. Croat emigres were important in funnelling money from the USA, Canada and elsewhere. In the émigré communities, more than in Croatia itself, Ustasha ideology lived on, especially the belief that defending Croatian independence required armed struggle.[3] One could say that those Croats who did not want to be restructured under the Yugoslavian communist system emigrated from Yugoslavia, often with staunch nationalism in their habitus.

During 1990 and 1991, the Bosnian Serb areas had been provided with weapons by the Yugoslav Army, as part of the so-called RAM programme. The programme had been public knowledge since September 1991.[4] The leading force was the military line where Ratko Mladić played a role, but with Milošević's knowledge.[5] Mladić began to have increasing political leverage as the situation radicalised, which is noteworthy as he was no politician, but the restructured field opened up a new space of possibilities for him. According to Louis Sell, the political takeover of Tudjman's HDZ regime in Croatia was a major factor for this: "But it was Tudjman's HDZ regime, which the JNA viewed as a modern reincarnation of the murderous Ustasha that really made the generals see red".[6] Further, weapon smugglers had travelled between all parties in Bosnia in the year before the outbreak of war selling weapons under the pretext that weapons had been sold to the other parties.[7] It is alleged that this weapon smuggling extended right into the Bosnian Parliament. It was therefore not a question of small-scale traders flogging what they could, but of weapons deals in which members of the republic's Parliament were involved, partly the same people who

---

[2] Gagnon (2004, p. 103), Mann (2005, p. 381), Naimark (2001, p. 170), see also Donia and Fine (1994, p. 210), although without exact dates and places. [Tribunal update 68. Stipe Mesic's testimony, 16–21 March 1998].

[3] Mann (2005, p. 377).

[4] Udovicki and Stitkovac (1997, p. 179).

[5] Sell (2002, p. 123), Mann (2005, p. 390).

[6] Sell (2002, p. 122).

[7] Udovicki and Stitkovac (1997, p. 180).

had been responsible for throwing the country headlong into war.[8] It was therefore common knowledge, more than half a year before the declaration of independence, that the Bosnian Serbs in particular were heavily armed.

Early in 1991, Milošević and the Slovenian leader Milan Kučan declared each nation's right to follow its own path, an agreement which put Croatia in a difficult position as they were not a part of the agreement but bordered to both of the countries—or rather both parts of Yugoslavia.[9] On 25 June 1991, Slovenia and Croatia declared their independence from the Yugoslav Federation.[10] This in itself was the manifest defeat of Yugoslavia as a political system and also for the rules of the social field of politics in the geographical arena of ex-Yugoslavia. Earlier that same month in a statement made in Belgrade, the US Secretary of State, James Baker, had announced support for a united Yugoslavia.[11] EU countries such as Austria, Germany, Hungary and Denmark actively supported Slovenia's and Croatia's efforts to gain independence during the spring and summer of 1991.[12] France and Great Britain maintained a more reserved stance, so the EU was far from united on the issue—but in the end it was the active, positive element, not the passive, more muted group that won. The parties thus received different messages from the EU and the USA, which conferred legitimacy for both camps, separatist and federalist alike. The ten-day-long war in Slovenia ended on 8 July 1991.[13] Serbia had no border with Slovenia and neither were there any Serbian minorities in the republic, both factors contributing to Slovenia coming out of the conflict relatively unscathed. Added to this was the fact that the Slovenian forces were relatively strong for a small constituent republic. The absence of Serbs in Slovenia was also something that prevented Milošević from claiming areas of the republic,

---

[8] "Rovosi u dusi" Zehrudin Isakovic. *Vreme* 911216, p. 24.

[9] Sell (2002, p. 128).

[10] Donia and Fine (1994, p. 218).

[11] Donia and Fine (1994, p. 220).

[12] Woodward (1997, p. 219). It should be noted here that the split within the EU was quite marked, with countries like France and Great Britain opposing the separatist line. Ibid., p. 223.

[13] Stitkovac (1997, p. 159).

but overall there was a great deal that indicated that a war with Slovenia would not be a particularly successful venture.

The Brioni Accord signed on 7 July 1991 was a form of armistice after the short Slovenian War. The accord was sanctioned by a significant part of the international community, including the EU and the USA. The agreement recognised Slovenia's independence. This, however, invalidated the legitimacy of all those within the Yugoslav Army who had been willing to take up the cause of Yugoslav unity. There was now no longer either international or national support for the idea. Slovenia's independence rendered Yugoslavia an army that now had separatism, bloody or non-violent, as the only alternative.[14] It so happened, however, that the strongest separatist forces in Croatia and Serbia were not interested in peaceful solutions, which was also no secret to anyone. The army was thus driven into the hands of the politicians who were willing to conduct their policies using violence to achieve their aims in line with their *habitus*. To describe the political actions of the countries involved as unwise and showing lack of judgement would be an understatement. Milošević was able to exploit this effect to strengthen his grip on the Yugoslav Army. On his part, it demonstrated a skilful exploitation of the actions of other actors, and his knowledge of the local field was overwhelmingly the same as that of the international community.

The Badinter Commission was formed in August 1991, when it became clear that in one way or another Yugoslavia would become partitioned. The Commission's purpose was to ensure this happened in the most fitting manner. It directed that an official referendum should be held in Bosnia, ensuring that the three main ethnic groups should be strongly represented in the voting process.[15] The intention of the EU and the Commission was questioned by the Bosnian Serbs. From a Bosnian Serbian perspective were the EU not even an actor on the field and thus lacked field specific capital, which is needed for legitimacy. That the EU disposed of other types of capital was of course clear but power is certainly not always followed by legitimacy.

---

[14]Woodward (1997, p. 223). See also Sell (2002, p. 146).

[15]Donia and Fine (1994, p. 238). See also Sell (2002, p. 163).

As it has been pointed out by Mann, "the biggest opposition parties were even more nationalist than Milošević".[16] This is true and that's the reason why Milošević could go as far as he could and still win a lot of the confidence of the people—he didn't appear all to extreme even if he went in that direction—the social field was tilting towards a new logic of practice. Still, the blame was still very much on Milošević. He was disposed for opportunism. There was little reason for him to go against the grain and oppose the radical currents which one instead could pick up on and use to build power on. Mann also points out that pre-election surveys showed that important issues for the people was the communist legacy, the economy, living standards, good international relations but also the defense of the nation.[17] The political agenda of the voters looked rather normal for a civil society, at least more civil than the agendas of the parties in Serbia. What can be said is that the demand of defence of the nation—which is perfectly in order to demand—got a dark side in that the definition of the borders westward was a part of the contemporary political discussion. It is not given that defence of the nation is a defensive stance. Despite that Milošević by far had the strongest position on the political field, he did not use it to moderate the political climate; instead, he did go with the flow of the field as a true opportunist in order to maximise his influence.

Another political actor who temporarily rose to some power, Vojislav Šešelj, had experimented with founding a party the year before, 1991. He did form SRP (*Stranka Srpskog Jedinstva; The Party of Serbian Unity*) in 23 February 1991.[18] Šešelj did among a host of other things threaten Croats in Vojvodina with expulsion and confiscations and that in no less prominent arena than in the Serbian Parliament.[19] This was certainly a sign of a new political practice in coming. SRP, under the leadership of Vojislav Šešelj, did from its start work for the dissolution of Yugoslavia and a strong Serbian state with Serbia, Montenegro and Krajina within its borders according to the Karlobag–Karlovac–Virovitica formula.[20] From its start, the SRP did form paramilitary units, as a part of its Chetnik modus operandi. They did first fight in the Croatian war and later in Bosnia-Hercegovina.[21]

---

[16] Mann (2005, p. 371).

[17] Mann (2005, p. 372).

[18] Thomas (1999, p. x).

[19] Kerenji (2005, p. 376).

[20] Stojanovic (2000, p. 465).

[21] Stojanovic (2000, p. 470).

In August, war broke out between Serbia and Croatia. The sacking of the city Vukovar was a brutal affair with high causalities on all sides. The JNA did expectedly side with Serbia, or in other words: "The war in Croatia fully revealed the teaming up of the Serbian and the army leaderships, and turned JNA into an instrument of the Serbian regime's policy".[22] It is important to note that the army was very much associated with the communist party, which Milošević was the heir of. For example had JNA a representation of its own in *The League of Communists of Yugoslavia*, which only the Yugoslav republics had representation in.[23] The army was thus integrated into the politics and had power at stake; if the system dissolved, then the army would lose power: "The party domination over the army resulted in the ideological organisation of the JNA, and, accordingly of the whole defence system".[24] Still, despite the army's intentions, it didn't exercise enough control to actually have its soldiers to turn up, which also is an indicator of the lack or popular support in Serbia for a war in Croatia. The Yugoslav Army was short of 18 divisions at the start of the Croatian War. The shortage was the result of desertion and a refusal to report for military service. TV meanwhile served to legitimise local nationalism and blinker out moderating opinion.[25] In fact, 50–85% of Serbs called up to fight in Croatia didn't show up.[26] This is a circumstantial evidence that the war in Croatia but also to an extent in Bosnia was more of a top-down war than a bottom-up war.[27] But then again, few wars are forced upon the political leadership by the population. Knowledge of this could have made psychological operations of value at this stage, certainly as it is not the question of direct military intervention which would have been impossible at this stage. One can always discuss the value of PSYOPS, but in the end it is an empirical question if it is of value on the given situation or not.

The sacking of Vukovar in Eastern Croatia did also include paramilitary forces, not only the army. The most notorious was *Arkan's Tigers*. Arkan was the leader of the football hooligans of the Red Star. These

---

[22] Hadzic (2000, p. 527).

[23] Pesic (1996, p. 44).

[24] Hadzic (2000, p. 514).

[25] For example, Miloševic (1997, p. 108). For more on the recruiting problem: Sikavica (1997, p. 142).

[26] Gagnon (2004, p. 109).

[27] Gagnon (2004, p. 51).

hooligans were the core of his "Tigers".[28] Arkan had personal contact with Milošević, and Arkan made no secret of his contacts with the state security in Serbia.[29] The Cetniks under Šešelj also had their own force, and Šešelj made gory statements as: "We must cut the Croats' throats, not with a knife but a rusty spoon".[30] This complemented with the Chetnik paramilitary group White Eagle's leader Mirko Jovic who said, "We are not only interested in Serbia but in a Christian, Orthodox Serbia, with no mosques or unbelievers [...] I am all for the clearing operations".[31] The Chetniks were armed by the JNA and Arkan's Tigers by the Ministry of Interior of Serbia.[32] Vuk Drašković' SPO also had a paramilitary group the "Serbian Guard" which was formed in 1991 and also saw combat.[33] Ironically, Croatian Police captured Arkan in November 1990 in Croatia but happened to release him to Belgrade in June 1991, just before the war started.[34] Fighting did not only occur in eastern Croatia, but also in the Serb-dominated Krajina in Croatia. Croatia lost control of Krajina between June and December 1991. The intention for the local Serbs was to make it a part of the reformed Yugoslavia.[35]

Acting fast, Germany formally recognised Croatia on 23 December 1991. Neither the EU as an institution nor Germany individually did ensure that Croatia held to the guarantees made to the minorities living within Croatia's boundaries.[36] This was a signal to the Serbian minorities in Croatia—but also in Bosnia—that they would have to rely on the remnants of Yugoslavia, i.e. Serbia, rather than the international community for their security as violence already had been shown in full force in other parts of the now defunct state of Yugoslavia. This threatening situation for Serbs was what Milošević had warned against earlier. He had been shown right—despite the fact that his reasoning had no

---

[28] Colovic (1996, p. 386).

[29] Miljkovic and Hoare (2005, p. 205). Pavlakovic' (2005, p. 22).

[30] Mann (2005, p. 392).

[31] Cited in Mann (2005, p. 392).

[32] Mann (2005, p. 392).

[33] Stojanovic (2000, p. 475). 12/13 March Draskovic released from prison. Thomas, p. x.
[34] Gagnon (2004, p. 147).

[35] Lukic' (2005, p. 55). The conflict in Krajina in 1992 did result in an estimate of 3000–6000 deaths. Tabeau and Bijak (2005, p. 198).

[36] Woodward (1997, p. 226).

realistic foundation at the time but rather was propagandistic. By failing to put pressure on Croatia, the EU managed once again to strengthen the legitimacy of Milošević's policies and his position on the political field of Serbia in general (or parts of a Yugoslavian political field if one wants to define the social space as such). This was a legitimacy that he had previously lacked among many Serbs who earlier had been uncertain about him. This was also a decisive blow to the Serbian politicians who still sought a peaceful agenda by constitutional civil political means—they found themselves stripped of their legitimacy as a result of having argued against Milošević's earlier preaching that the international community was against the Serbs. On the other hand, the war was polarising and popular support was not necessarily strengthened by making war on neighbours. Unsurprisingly, the war reduced Milošević's popularity. Faced by public opposition, in 1991 (and also 1993 due to Bosnia) he resorted to coercion. His formidable police powers ultimately swept demonstrators off the streets and closed down independent media on trumped-up charges.[37] Milošević marginalised and tried to silence the opposition in 1991.[38] One can discuss the media's role when it came to unleashing the more grievous events in the breakup of Yugoslavia. A lot of the papers, as *Vreme*, had a rather balanced view and some got in trouble for its criticism against Milošević. But when one looks at a social field, one should bear in mind that the vehicles of information differ between different social groups but also within general social classes, their way of distinction differs. *TV Novosti* was a weekly Serbian paper with the middle class and lower class as primary consumers. On 12 July 1991, one could read the following in it: "Concealed by the so-called 'Brioni Declaration', which in fact simply froze the Yugoslav Army disaster in Slovenia while obliging the army in Croatia to withdraw to barracks, leaving the Serb inhabited areas at the mercy of the new pro-Ustasha government—whose genocide intentions could not be doubted—it is hardly necessary to draw the parallel with the Yugoslav catastrophe of April 1941".[39] The reference to the Second World War

---

[37] Mann (2005, p. 373).

[38] Gagnon (2004, p. 103).

[39] Cited in Markovic (2000, p. 605).

is a clear way of giving legitimacy to a construction of history which prescribed a violent political agenda. Another example is *Illustronova Politika* which was a weekly Serbian paper with the middle class as target. On 30 July 1991, one could read in it: "Our motive is not to allow a repeat of 1941 when the Ustasha, the ancestors of today's HDZ, massacred the people here".[40] Even here the references are clear: do not be a bystander if your people are harassed once again. Historical references were used like this to tap further capital into the agendas of those with political power.

Others reacted in other ways. The rather belligerent and nationalistic Drašković changed his agenda very much after the sacking of Vukovar.[41] Drašković lost influence as he turned his political agenda around; he was not convincing as a liberal and old hardliners left.[42] His habitus was not structured for such a turn. Drašković transformed as a politician and wanted peaceful means as political practice. Still, he wanted Croatia to cede areas both to Serbia and to BiH (in the latter case, it would be the ceding of Krajina from Croatia) (Fig. 7.1).[43]

It should be emphasised that the dangerous political process described above occurred just before the war and sometime into it. The question of guilt for this process is not the most relevant, rather *how is it that X happens*. Regardless of intent, one must say that Tudjman, Milošević and many other actors were not status quo actors. Tudjman acted on behalf of his nationalism and Milošević on his own behalf.[44] There were as seen structural reasons for things to happen, but one must take both actors and structures into account to understand change in society. Some researchers have seen a security concern as the root of much of the violence in Bosnia.[45] There is some validity in that statement—as it puts emphasis on certain practices and also tries to go for a structural explanation in a reasonable way; even if it does not tell the complete story, not all security concerns result in war. With the changing structure after

---

[40] Cited in Markovic (2000, p. 606).

[41] Mann (2005, p. 374).

[42] Stojanovic (2000, p. 473).

[43] Stojanovic (2000, pp. 463, 474).

[44] For an argument about Tudjman and Milosevic not being status quo actors, but from a different perspective, see. Roe (2000, p. 386).

[45] Dulić and Kostic (2010, p. 1067). I disagree that the events unfolding necessarily *needed* a perceived threat, but I do think that there was such a perception and that it contributed to the events. Ibid., p. 1069. There were more factors in play than just threats for these events to unfold.

The fall of structuring structure of the Cold War opens up new spaces of possibilities in the Yugoslavian political field

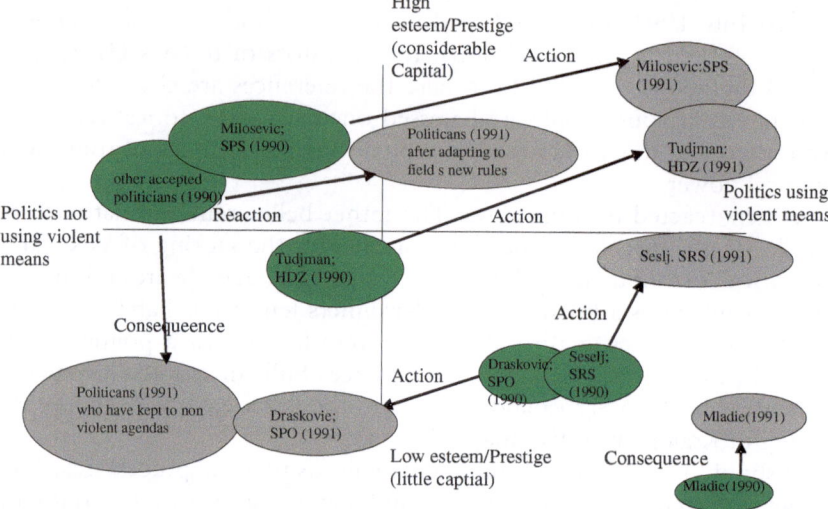

Fig. 7.1 A mind map of the political field of the crumbling Yugoslavia with some of its actors. The field above shows the positions of a number of key political actors on two different occasions. As a result of capital strong actors sanctioning violence as a political means, this method gains legitimacy in the political field

the Cold War, new possibilities opened up and some of the actors had a social disposition which eventually led to the political field changing and hostilities breaking out. There is thus little chance of influencing the process other than with firm, clear and forceful diplomacy. Sending troops in that phase could only have been done with Yugoslavian approval, and was something neither Tudjman nor Milošević would ever have sanctioned. The actual situation at the time was that there were no substantial ground forces available to send from either the EU or the USA, even if they had been given a hypothetical green light to deploy to Yugoslavia. The political field did not provide any viable alternative; when Yugoslavia begun to break up, Milošević's "all Serbs in one state" had resonance.[46]

---

[46] Stojanovic (2000, p. 466).

The scene was set both in Croatia and also in Bosnia for a violent near future. This was not very hard to perceive taken the turns the politicians in the former Yugoslavia had taken and the recent violent logic of practice of the same politicians.

BIBLIOGRAPHY

Colovic, I. (1996). Fudbal, huligani I rat. In Nebojsa Popov (Ed.). *Srpska streana rata* (pp. 435–444). Belgrad: Republika.

Donia, R., & Fine, J. (1994). *Bosnia and Herzegovina. A tradition betrayed.* London: Colombia University Press.

Dulić, T., & Kostic, R. (2010). Yugoslavs in arms: Guerilla tradition, total defence and the ethnic security dilemma. *Europe-Asia Studies, 62,* 7 (New York: Routledge).

Gagnon, V. P. (2004). *The myth of the ethnic war. Serbia and Croatia in the 1990s.* Ithaca: Cornell University Press.

Hadzic, M. (2000). The army's use of trauma. In N. Popov (Ed.), *The road to war in Serbia. Trauma and catharsis.* Budapest: Central European University Press.

Kerenji, E. (2005). Vojvodina since 1988. In S. P. Ramet & V. Pavlakovic (Eds.), *Serbia since 1989. Politics and society under Milošević and after.* Seattle: University of Washington Press.

Lukic´, R. (2005). From the federal republic of Yugoslavia to the union of Serbia and Montenegro. In S. P. Ramet & V. Pavlakovic (Eds.), *Serbia since 1989. Politics and society under Milošević and after.* Seattle: University of Washington Press.

Mann, M. (2005). *The dark side of democracy. Explaining ethnic cleansing.* Cambridge: Cambridge University Press.

Markovic, Z. M. (2000). The nation: Victim and vengeance. In N. Popov (Ed.), *The road to war in Serbia. Trauma and catharsis.* Budapest: Central European University Press.

Miljkovic, M., & Hoare, M. A. (2005). Crime and the economy under Milošević and His successors. In S. P. Ramet & V. Pavlakovic (Eds.), *Serbia since 1989. Politics and society under Milošević and after.* Seattle: University of Washington Press.

Milošević, M. (1997). The media wars. In J. Udovicki & J. Ridgeway (Eds.), *Burn this house. The making and unmaking of Yugoslavia.* Durham: Duke University Press.

Naimark, N. M. (2001). *Fires of hatred. Ethnic cleansing in twentieth-century Europe.* Boston: Presidents and Fellows of Harvard College.

Pavlakovic, V. (2005). Serbia transformed? Political dynamics in the Milošević Era and after. In S. P. Ramet & V. Pavlakovic (Eds.), *Serbia since 1989. Politics and society under Milošević and after.* Seattle: University of Washington Press.

Pesic, V. (1996). Serbian Nationalism and the Origins of the Yugoslav Crisis.*Peaceworks 8, 31,* 12. Washington D.C: United States Institute of Peace.

Roe, P. (2000). Former Yugoslavia: The security dilemma that never was? *European Journal of Relations, 6,* 373 (London: Sage).

Sell, L. (2002). *Slobodan Milošević and the destruction of Yugoslavia.* Durham: Duke University Press.

Sikavica, S. (1997). The army's collapse. In J. Udovicki & J. Ridgeway (Eds.), *Burn this house. The making and unmaking of Yugoslavia.* Durham: Duke University Press.

Stitkovac, E. (1997). Croatia. The first war. In J. Udovicki & J. Ridgeway (Eds.), *Burn this house. The making and unmaking of Yugoslavia.* Durham: Duke University Press.

Stojanovic, D. (2000). The traumatic circle of the Serbian opposition. In N. Popov (Ed.), *The road to war in Serbia. Trauma and catharsis.* Budapest: Central European University Press.

Tabeau, E., & Bijak, J. (2005). War related deaths in the 1992–1995 armed conflicts in Bosnia and Herzegovina: A critique of previous estimates and recent results. *European Journal of Population, 21,* 2. Heidelberg: Springer.

Thomas, R. (1999). *Serbia under Milošević politics in the 1990s.* London: C Hurst & Co Publishers Ltd.

Udovicki, J., & Stitkovac, E. (1997). Bosnia and Hercegovina: The second war. In J. Udovicki & J. Ridgeway (Eds.), *Burn this house. The making and unmaking of Yugoslavia.* Durham: Duke University Press.

Woodward, S. L. (1997). International aspects of the war. In J. Udovicki & J. Ridgeway (Eds.), *Burn this house. The making and unmaking of Yugoslavia.* Durham: Duke University Press.

CHAPTER 8

The Events in 1992

Abstract This chapter deals with how and why the war emerges in Bosnia and how the involved actors acted. Serbian and Croatian leadership with Slobodan Milošević and Radovan Karadžić, respectively, pushed for hostilities all when the leadership in Bosnia with Alija Izetbegović and parts of the international community as Germany did little to avert the coming disaster.

Keywords Bosnia · Serbia · Croatia · Alija Izetbegović

As a result of mainly German pressure, the EU followed Germany's action and recognised Slovenia and Croatia on 15 January 1992, without any guarantees being given to the Serbs living in Croatia.[1] The symbolic significance of the event was apparently paid such respect that no actual countermeasures in the event of conflict were prepared. This recognition was as questionable as Germany's, seeing that Croatia was still not being governed in a manner sufficiently satisfactory to preclude internal unrest, particularly bearing in mind the problems inherent in the Krajina region. One should also remember the lack of guarantee to the minorities in

---

[1] Donia and Fine (1994, p. 233).

© The Author(s) 2017
H. Gunneriusson, *Bordieuan Field Theory as an Instrument for Military Operational Analysis*, New Security Challenges, DOI 10.1007/978-3-319-65352-5_8

Croatia mentioned earlier; therefore, the Croatian Serbs could not expect the support of the international community. They were thus forced to seek help from other Serbs, something which the Milošević-controlled Yugoslav Army had provided in the form of weapons. For the Serbs cut off in Croatia, Milošević was therefore seen as a man of vision, a defender. To oppose his policies appeared more and more pointless, especially when through his actions and as a symbolic figure, he was supported by a number of actors who did not necessarily have it in mind to help Milošević. In this way, the impression that the move of Milošević to the right on the field was justified was confirmed in practice, and he was now one step ahead. That Milošević's subsequent manipulative manner of introducing violence as a political means can be considered shameful, which is not relevant here. In the local arena, a number of reasonably well-considered moves at the time by Croatia and other countries served to legitimise a shift to the left of the political field.

In this phase, the events displayed in Fig. 7.1 take on a more incisive aspect, with politicians inclined towards constitutional means losing capital in the political field; alternatively, they alter their agenda and move to the right side of the field when they realise that the policies that have polarised on the right are successful. Fighting between Serbs and Croats immediately after Croatia's declaration of independence took place mainly in the Serb-dominated area of the city Knin (*Kninska Krajina*) in Croatian Dalmatia and also in Slavonia. In January 1992, a ceasefire was agreed upon which then came into force. This opportunity was exploited by transferring troops to Bosnia instead, or, as it might be described, utilising what in clausewitzian terms is called inner lines of communication. Bosnia's president, *Izetbegović*, failed on Bosnia's behalf to follow up the situation or appreciate the seriousness of disquieting reports of military exercises in Bosnia.[2]

From a military viewpoint, Bosnia was special, with the majority of the Yugoslav defence industry being based there. In addition, Bosnia throughout history, and especially during the partisan fighting of the Second World War, had always been important for the defence of the region. The mountainous and forested terrain of the region made it easy to defend, and it was therefore here that most of the Yugoslav military mobilisation stocks were held; it was planned that the area would act as a

---

[2] Udovicki and Stitkovac (1997, p. 183).

military fulcrum in the event of invasion.[3] From a doctrinal perspective, giving up Bosnia would be tantamount to surrendering Yugoslavia. This appreciation of Bosnia in the military mind has not been brought out in literature, but it is an important factor if one is to understand the scenario that came to be played out in that constituent republic.

The Badinter Commission's intention to get the referendum in Bosnia as the civil way dealing with Bosnia's partition from Yugoslavia did not occur. The Bosnian Serbs essentially boycotted the voting for Bosnia's independence on 29 February 1992.[4] In January and February, when they could have taken the political initiative, the EU was waiting for the outcome of a referendum that the Bosnian Serbs had already said they would boycott.[5] The declaration of the boycott presented an ideal opportunity for the EU to get inside the decision cycles of the parties involved, but instead, they chose to ignore the information. The EU recognised Bosnia on April 6, and USA followed suit the next day.[6] The Bosnian Serbs' boycott of the referendum was not recorded, and the recognition of the Bosnian state—without the question of the Bosnian Serb position being even discussed—served to legitimise the theory of the Bosnian Serb politician, Radovan Karadžić, that the outside world was against the Serbs.[7] The trigger for the war in Bosnia was the EU's recognition of the Bosnian state on 6 April 1992.

All was not peaceful in Croatia either. Karadžić became radicalised during the war, playing more and more on racism and cultural differences in his speeches, visiting Orthodox masses to a greater extent, increasingly positive to hostage taking and generally having an uncompromising attitude.[8] Karadžić, however, was an actor in the political field

---

[3] Donia and Fine (1994, pp. 155 and 174). There was an idea of central defence, striking out of Bosnia. Not dissimilar to Sweden in the 1800s when Sweden's defence policy was based on defence of the central area of Sweden to strike out from. Bosnia's importance for the defence of Yugoslavia can be compared to that of Karlsborg Sweden, and its surroundings.

[4] Donia and Fine (1994, p. 238). See also Sell (2002, p. 163).

[5] Udovicki and Stitkovac (1997, p. 231).

[6] Donia and Fine (1994, p. 236). The authors believe that the USA led this process.

[7] Udovicki and Stitkovac (1997, p. 178).

[8] Owen (1995, p. 301).

by virtue of previously having been a politician, an actor with political capital from politics as a parliamentary practice. Thus, when comparing, for example, Karadžić with Mladić, one finds that these two actors acquired their legitimacy in the political field from different forms of capital, which meant that they could also be influenced in ways that themselves were different. But to do that one must be aware of the mechanics of the field as well as aware of the types of capital the named actors stockpiled. Worthy of mention is that opinion polls reported in the paper *Vreme* in 1992 showed a slide in SPS popularity and by that also the popularity of Milošević. This was especially true in the cities, in the north of Serbia, and among the educated. Only 15% of Belgraders supported the SPS compared to 51% of those in south Serbia.[9] Milošević no longer held a strong standing among the Serbs in general.

The Bosnian Serb General Mladić, standing as a political force, was greatly improved in power by this change on the field; political power came into his reach as the military power he disposed became legitimate political tools. Mladić's authority as a politician was to a large extent founded on the use of violence as a political means. If that approach became difficult to adopt in the political arena, then Mladić's position would also be undermined. This was something that he himself certainly did not have any problem appreciating, even if that appreciation was not conducted in theoretical terms. In the picture above, Mladić could be replaced with other high-ranking military officers who started to play a political role, ones the field had changed. Sir Rupert Smith describes that Mladić's political power came from the weapons at his disposal, which is in line with the theory described.[10] Smith also describes how attacks on, for example, Mladić's home village were made in order to undermine the symbolic capital which Mladić had and thereby undermine his power.[11]

When Bosnia was recognised as a state on 6 April 1992, all of a sudden the Yugoslav Army became a foreign army in the region. This happened because the real point at issue had not been debated at an

---

[9] Mann (2005, p. 373).

[10] Smith (2005, p. 367). See also Owen (1995). He describes Mladić as a man conducting battles partly like an intellectual, partly like a barbarian, p. 280.

[11] Smith (2005, p. 366).

international level, and statements had just been made without further analysis. Proposals from the new government of Bosnia to withdraw were met by the army's reply that 80–90% of the troops stationed in the area were Bosnian and had no intention of leaving the country.[12] At the end of April 1992, the Bosnian Serbs took control of the Banja-Luka area in Bosnia and almost immediately started using the former local Yugoslav TV substation to spread war propaganda.[13] A duel was conducted during the war in Bosnia between the TV channels of the different (political) camps, and a mortar attack on a bread queue in Sarajevo, for instance, being given a different slant by both TV Sarajevo and TV Pale.[14] It is also worth noting that Bosnian TV was also used as an instrument of propaganda, promoting the idea, for example, that the war was being waged by people who did not live in Bosnia. This was a lie, even if volunteers from mainly Croatia and Serbia were represented among the fighting elements.[15] In May 1992, 38 generals were purged from the JNA to establish Serbian control.[16] The Bosnian politicians driving forward the move towards independence should most certainly have understood that this situation would arise, and they were playing a dangerous game. The Bosnian President, and the leader of the Bosnian Muslim Party, Alija Izetbegović, chose to declare Bosnia independent, inspired by the example set by Croatia, despite other Bosnian Muslim voices warning against it.[17] Izetbegović apparently misjudged the situation in a grave way. It is also relevant that the Bosnian Serb leader, Karadžić, had not been at all unclear on the issue of what Bosnia could expect if it happened to declare independence.[18] The situation in Bosnia was much different compared to Croatia and Slovenia, as Robert Hayden discusses.[19]

---

[12] Sikavica (1997, p. 146).

[13] Udovicki and Stitkovac (1997, p. 187).

[14] Udovicki and Torov (1997, p. 121).

[15] Loc. cit.

[16] Mann (2005, p. 394).

[17] Udovicki and Stitkovac (1997, p. 175) [Nadezda Gace, "Velika Srbija na mala vrata" *Vreme* 911127, p. 27].

[18] Udovicki and Stitkovac (1997, p. 179) [Roksanda Nincic et al., "Drina bez Cuprije" *Vreme* (weekly anti-war publication, Belgrade), 911021, p. 20].

[19] Hayden (1992, p. 661).

The new Bosnian state did not control most of its territory, which really should not come as a surprise to any of those who recognised the state. That knowledge combined with the recent acts of violence in former Yugoslavia should have been enough of a warning of the events to unfold.

Even though David Owen had good experience of Alija Izetbegović, he did not hesitate to mention that this opinion of Izetbegović was not shared by others. Manipulative and untrustworthy are labels which were used. But his indecision, which has been mentioned above, is also mentioned along with fundamentalistic advisors.[20] Owen describes Izetbegović's party SDA as increasingly intolerant, which fits very well into the theoretical picture above (Fig. 7.1). In order to keep being a political factor, one has an easy option to allow oneself to be radicalised by the conflict.[21] One can also note that the Bosnian Muslims did not identify themselves very much as Muslims before the war, and the genocide of them changed that.[22] The logic of practice changed the perception of the self so that the Bosnian Muslims restructured themselves. The former deputy of Carl Bildt—the first high representative for Bosnian peace implementation—Louis Sell confirms this with stating that Izetbegović called for a creation of a Muslim federation from Indonesia to Morocco and that media only should be entrusted to people of "deeply Islamic faith".[23]

Izetbegović's not being able to judge the situation correctly may perhaps be explained by the fact that he had poor contact with the Bosnian people outside Sarajevo, a city that was to a marked degree multi-ethnic and not representative of the rest of Bosnia. In any event, he placed far too much hope on the international community's will and ability to intervene. In addition, he had the task of representing different sections of Muslim political persuasion, something that contributed further to the image of him as indecisive.[24] Owen thinks that Sarajevo was besieged by both Serbs and the Bosniacs, who got political leverage by having the Serbs continue besieging the city. People were not allowed to leave the

---

[20] Owen (1995, p. 38).

[21] Owen (1995, p. 40). See also Sell (2002, p. 5).

[22] Gagnon (2004, p. 27).

[23] Sell (2002, p. 158).

[24] de Graaff (2003, p. 114).

city as the political positive effect for the besieged would dwindle by that.[25] There was also documentation of beleaguered Bosniacs drawing Serbian fire on a hospital by placing mortars behind it, and sniper fire on civilians in the city from a building controlled by the Bosnian Government.[26] This theory is plausible as it was a political decision, top down, to use the citizens of Sarajevo as a means of politics; it was not the people who put themselves in the line of fire. The Bosnian politics had hardly any other cards to play than the card of inciting the picture of suffering and lack of power, towards the international community, and this was one of a few viable ways to conduct international politics for Izetbegović and his allies.

This still did not ring enough alarm bells for the Bosnian and international politicians who chose to act, ignoring the statement by the Badinter Commission affirming the need for all three ethnic groups to be well-represented in order for the referendum to be considered valid. This was certainly evident when the Bosnian Muslim leader Izetbegović, under the influence of the USA, withdrew from the EU-supported Lisbon agreement in 28 March 1992, which would have divided Bosnia into cantons—both the Croats and Serbs agreed to this. The USA tolerated the Bosnian Muslims covertly importing weapons from Iran and other Muslim countries.[27] The Croatians, on the other hand, evaded the arms embargo (despite the ineffective resolution UNSCR 713) and equipped both themselves and the Bosniacs with arms as no UN surveillance existed between Posavina and Western Herzegovina.[28] The recently reunited Germany had no use for the old Eastern German munitions it had inherited, but Germany had a political agenda in Yugoslavia. Croatia had the opportunity, despite the embargo, to buy former DDR heavy artillery and tanks with other countries as middlemen. This of course stiffened Serbian resistance against demilitarisation, for they still had

---

[25] Owen (1995, p. 59). One can also note a sniper incident at Sarajevo, performed by Bosniacs, killing an UNPROFOR soldier on 8 September 1992. Apparently, it was done as an attempt to make it look like a Serbian attack, as the soldier was escorting food to the beleaguered Sarajevo. Owen (1995, p. 44).

[26] Owen (1995, p.106).

[27] Sell (2002, p. 224).

[28] Owen (1995, pp. 45, 47 and 315).

a weapon embargo.[29] The situation did a lot to further restructure the political field in the remaining Yugoslavian area as it was clear that the international community was more against Serbia than was neutral.

All of this is important for outsiders to understand if we in the future intend to intervene in situations that might involve the deployment of troops. The EU and the USA maybe found it more difficult to see the problems in Bosnia, especially when the separatist Bosnian politicians did not raise the issues at the international level. But more caution could most certainly have been exercised in the period before Bosnia declared independence. The signals from the Bosnian Serbs had been clear at the time of the referendum. There existed an imbalance in the capacity for violence that favoured the Bosnian Serb minority. This made a violent interpretation of the field more likely. An important part of the collective habitus in the scenario is Yugoslavia's violent recent past with infighting and ethnic cleansing during World War II, where the capacity for violence became a legitimate form of political capital. In addition, the party system in Bosnia followed ethnic and not ideological patterns; the choice of nationalist party programme was geared to ethnic origin.[30] Tudjman has been described as having power over the political discourse and changing it into a hostile one regarding ethnic origin in particular. This in its turn could "enhance one's status in the community".[31] This can easily be seen as a social field where the actors with consecrating power restructure the field, and other actors either follow the new structure with their power intact or strengthened, or they resist and might eventually lose power as the new structure does not support their choice (a conscious choice or not).

The idea of ethnic background dictating dividing lines was nothing new for Bosnia as part of the former Yugoslavia, but it stands to reason that one can clearly expect problems in a republic that has aspirations of independence and that at the same time has ethnic diversity considered to be an inherent political issue by all major parties. Added to this, the neighbouring countries with the same ethnic diversity had waged a war

---

[29] Owen (1995, pp. 70 and 74). See also p. 120 for reference that even the Bosniacs had gathered an arsenal of quite some substance in 1995. In all honesty, the Serbs had a lot of the old Yugoslavian weapon production capacity under control, so they needed allies less than others in this respect.

[30] Udovicki and Stitkovac (1997, p. 175).

[31] Sekulic et al. (2006, p. 822).

along the same lines the year before. Given this scenario, there existed considerable potential for disaster.

The different opinions among the involved nations about what should be done were also a dilemma and a root for inaction and problems. The opinion in the USA was strongly anti-Serbian, which in its turn damaged the credibility of the peace brokers Cyrus Vance and David Owen and their attempts to negotiate a peace.[32] The Croats and Bosniacs received signals that they had an opportunity for a better deal by the USA than by Vance–Owen. The Serbs on the other hand saw that there was no united front among the European countries and certainly not in relation to the USA.

BIBLIOGRAPHY

Donia, R., & Fine, J. (1994). *Bosnia and Herzegovina. A tradition betrayed.* London: Colombia University Press.

Gagnon, V. P. (2004). *The myth of the ethnic war. Serbia and Croatia in the 1990s.* Ithaca: Cornell University Press.

Graaff de, B. (2003). The wars in former Yugoslavia in the 1990s: Bringing the state back in. In J. Ångström & I. Duyvesteyn (Eds.), *The nature of modern war: Clausewitz and his critics revisited.* Stockholm: Swedish National Defence College.

Hayden, R. (1992). Nationalism in the former Yugoslav Republics. *Slavic Review, 51*(4) (Cambridge).

Mann, M. (2005). *The dark side of democracy. Explaining ethnic cleansing.* Cambridge: Cambridge University Press.

Owen, D. (1995). *Balkan odyssey.* London: Harvest Book.

Sekulic, D., et al. (2006). Ethnic intolerance and ethnic conflict in the dissolution of Yugoslavia. *Ethnic and Racial Studies, 29,* 5 (New York: Routledge).

Sell, L. (2002). *Slobodan Milošević and the destruction of Yugoslavia.* Durham: Duke University Press.

Sikavica, S. (1997). The armys collapse. In J. Udovicki & J. Ridgeway (Eds.), *Burn this house. The making and unmaking of Yugoslavia.* Durham: Duke University Press.

Smith, R. (2005). *The utility of force. The art of war in the modern world.* London: Penguin Books.

---

[32] Owen (1995, pp. 100, 107, 136 and 295).

Udovicki, J., & Stitkovac, E. (1997). "Bosnia and Hercegovina: The Second War" burn this house. In J. Udovicki & J. Ridgeway (Eds.), *The Making and Unmaking of Yugoslavia*. Durham & London: Duke University Press.

Udovicki, J., & Torov, I. (1997). The interlude. In J. Udovicki & J. Ridgeway (Eds.), *Burn this house. The making and unmaking of Yugoslavia*. Durham: Duke University Press.

CHAPTER 9

Epilogue: Beyond 1992

Abstract The empirical example was to show just the initial years of the war, but a few things can be said of the last years of the war, without being conclusive on the matter. As war went on the violence escalated and the escalation gave way for even more violence. The logic of practice continued to change as the structuring field continued to be restructured. Without going into details it is worth mentioning that Milošević, who already had lost Serbian support, lost even more as he tried to balance between the international and national arenas at the same time. At the same time less international prolific actors could go full out on the national arena in support of Serbian interest.

Keywords The logic of practice · Milošević · Serbia

Vance-Owen had a plan which included keeping the Serbs separated. Mainly, Serbia represented by Milošević and Karadžić and Mladić on the other hand, representing the Republica Serpska, the Serbian part of Bosnia.[1] There was some success in this too, in August 1994, Owen noted that Karadžić and Milošević competed about being the foremost

---

[1] Owen (1995, pp. 105, 134,135, 143, 155, 158, 296 & 302. Especially, 155, 158, 296 & 302) show cracks between the leaders of Republica Serpska and Milosevic.

H. Gunneriusson, *Bordieuan Field Theory as an Instrument for Military Operational Analysis*, New Security Challenges, DOI 10.1007/978-3-319-65352-5_9

proponent for the Serbs.[2] This has similarities with the theoretical idea presented here. What the theory suggests is a stronger emphasis on identification of actors with capacity to political action without violence, and by that also identifies those lacking in that capacity. Trying to get Mladić away from the political field should then have been paramount, probably at the expense of giving both Milošević and Karadžićs more influence in the peace process than they had.[3] But in hindsight occurrences as the drawn-out siege of Sarajevo could have been shortened and the genocide in Srebrenica could have been avoided.

In addition to Bosnia's internal problems, one must also consider the situation in Yugoslavia as a whole. Despite the fact that the various ethnic groups in Bosnia appeared to get on well together—at least there had been no outbreaks of violence—the two leaders of Serbia and Croatia, Milošević and Tudjman had, as has been mentioned, met in 1991 and decided to split Bosnia between them. The agreement was thus struck two months after Croatia and Serbia had declared an armistice, showing that it was calculating statesmanship rather than irreconcilable nationalism that was guiding the two politicians. The two had already taken steps to prepare for military operations in Bosnia.[4] In addition, small elements of the Serbian and Croatian forces in Bosnia had moved on from the initial phase of murders committed by just small elite groups

---

[2] Owen (1995, p. 302). See also Sell (2002, p. 230).

[3] I am not saying that efforts were not made to separate Karadic and Mladić, but apparently they did not succeed that well. Owen (1995, p. 352).

[4] Donia and Fine (1994, p. 227). The authors highlight the infiltration of both Serbian troops from Krajina and Croat internal forces. Their views on another matter are unclear, because they are somewhat ambiguous. They refer to the fact that the Yugoslav Army used Bosnia for their operations in Croatia. Firstly, the war with Croatia was already over by the time of the Bosnian crisis, secondly—if they still claim that there were activities during the Croatian War which affected things in some uncertain way—the Yugoslav Army was for quite natural reasons and fully legitimately in Bosnia. It is difficult to imagine that the Yugoslav Army could have acted in any other way, but it is clear that the tension in Bosnia could have been increased as a result of all this, regardless of the right of the Yugoslav Army to be in Bosnia in 1991. This is particularly relevant considering that the Brioni Agreement made the idea of a united Yugoslavia an untenable strategy for the Yugoslav Army to align itself with. It was thus a time of uncertainty for the Yugoslav Army, not knowing to which country they belonged or whose army they were.

of paramilitaries and were now more readily disposed to carry out ethnic cleansing. For example, Vojislav Šešelj's party SRS had troops in Bosnia.[5] The stage was set for an exceptionally brutal and bloody war that could well have been avoided. In the end, the Serbs in Croatia did pay a severe price, or as Rupert Smith puts it:

The ethnic cleansing of the Croatian Serbs from Croatia was a prime example of the dynamics of the "theatre of war". While recorded and displayed at the time, the act was never attacked in the media for what it really was: the expulsion of a minority by a state from their homes on the basis of their ethnicity, and the failure of the UN to protect them, particularly as this, was the original purpose of the UN deployment.[6]

The very occurrence of starvation, ethnic cleansing and other active efforts of the participants also became a tool not only to rewrite the ethnic map of Yugoslavia, but also to affect the international community. "So ethnic conflict in Yugoslavia was old, but neither ancient nor continuous; and though it intensified in the 20th century, it did so unevenly".[7] Rupert Smith became aware of that during his stint as a commander in Bosnia and he confirms that the researchers Jan Willem Honig and Norbert Both were right in their statement:

The Muslim pockets were used by the [Bosniac] Sarajevo government in November (1992) as pressure points on the international community for firmer action. The longer that aid convoys were unable to reach them, the greater the pressure on the mandate. When convoys did succeed, calls for firmer action were unwarranted. Two weeks after the first successful delivery Muslims [Bosniacs] launched an offensive towards Bratunac [a Serb-held town just outside the besieged Srebrenica]. Thus the integrity of UNHCR and UNPROFOR was undermined, further convoys were impossible, and the pressure for firmer action resumed.[8]

What one can see from this is that—apart from the ruthlessness from the involved local parts—the violence was politicised and that prestige could

---

[5] Gagnon (2004, p. 50).

[6] Smith (2005, p. 362).

[7] Mann (2005, p. 359).

[8] Honig and Both (1996, p. 80), quoted in Smith (2005, p. 337). Owen notes that UN "knew" that Mladić would not honour any safe areas which were not demilitarised. Owen (1995, p. 67).

be won for all parts, even the international community, both through suffering but also through violence. Noteworthy is that the Commander in Chief of the Bosnian Serbian Drina Corps, General Krstić, said that none of the officers refused the orders in the Srebrenica massacre later on in the war in 1995.[9] By being a perpetrator one could also show that one was an important political player on the violent political field. This was especially important for actors like general Mladić, who had little political credibility beyond the political violence he could wield.

Rupert Smith has stated that "There was no strategic direction, there was no strategic military goal to achieve, there was no military campaign and there were no theatre-level military objectives: all acts were tactical".[10] That was a description of the planning for the international forces. If they had used a theoretical model similar to the one presented in this text, they might have been able to forge their tactical acts into some kind of operational and finally tactical pattern. The local factions were not equally unaware of the game which they all were involved in. Finally, NATO decided to act against Milošević and the Bosnian Serbian leaders and the UNPROFOR soldiers changed helmets from their blue to green—morphing into IFOR—while the bomb war started. This also gave the green light for the ethnic cleansing in Croatia of the Krajina. In early August 1995, about 150,000 Serbs were ethnically cleansed from Krajina, the largest single cleansing during the war.[11] The escalated use of violence as a legitimate political tool struck back at the Croatian Serbs, as the whole political field transformed into a social field where violence became the most effective way to wage politics—the Croatian state was no exception. One can compare it with the war related deaths in Bosnia during the war, 67,630 (minimum) or an estimate of 102,622 individuals.[12] The emigration from Bosnia (for whatever reason) as a result of the war is approximately 1.2 million individuals (27% of the population).[13]

The events following in Kosovo 1999 were also a major stroke against Serbia. It can serve as an example of how much symbolic capital

---

[9] Mann (2005, p. 396).

[10] Smith (2005, p. 336).

[11] Owen (1995, p. 329 & 353).

[12] Tabeau and Bijak (2005, pp. 203 & 207).

[13] Tabeau and Bijak (2005, p. 209).

Milošević's Serbia had lost internationally during the process of war. At that point, Serbia lacked so much symbolic capital that anything could pass as legitimate in order to weaken the state of Serbia. As the Independent International Commission on Kosovo states, "the NATO military intervention was illegal but legitimate" [sic].[14] In other words, there was enough symbolic value mustered to act against Serbia to make it legitimate despite formal writings. Russia was also weak during the bombings. Albeit they opposed the separations of Kosovo from Serbia, it did not mean much more at the time than spreading the symbolic violence to Russia as well. As we will see in the next chapter, Russia took notes in a way which the commission did not foresee in its conclusions.

Media deserves a special comment as media is powerful when it comes to affecting a social field in a society which is used to gather information by accessing media. Previously media has been mentioned as a means to affect diaspora or the population. The media and war have been interconnected as long as legitimacy has been a term of relevance. Modern thinkers such as Paul Virilio have taken the relationship of the media with war a stage further.[15] This will not be examined here, but it is just one more example of how theory can provide a perspective on and a valued explanation of war in empirical terms. In November 1994 when the war in Bosnia was creating bad will for Milošević, he shut down the newspaper *Borba* after it was critical of Milošević.[16] Repression against autonomous media is not unheard of and there are reasons for it as it affects the discourse on the events in a way not always benevolent to those in power. Another example of Milošević's awareness of the use of media was when he asked Carl Bildt for permission to use television against Karadžić, at the time when an agreement between Bildt and Milošević was made and a rift between the former and Karadžić was created.[17] Another example of the media war was when the authorities were still sending TV propaganda as late as March 1996 when the Serbs were driven out of Sarajevo.[18]

---

[14] The Independent International Commission on Kosovo (2000, p. 4).

[15] See in particular Paul Virilio's book on the Gulf War: *Desert Screen*.

[16] Pavlakovic´ (2005, p. 21).

[17] Sell (2002, p. 229).

[18] Udovicki and Stitkovac (1997, p. 200).

Television is an excellent means for the conduct of PSYOPS, especially if the warring parties and the local population both produce and watch TV. Clumsy efforts to conduct a media war were occasionally made by the international forces, for example by the USA through IFOR. This comment refers specifically to the fact that the TV media was not utilised constructively to any great degree by IFOR; that not much control of programme content was exercised, and that undesirable programmes could have been cut out when they appeared. In the case of 1996 above, IFOR was present on the ground and had the ability to act, at least in the media field. Instead, at first, IFOR concentrated on radio channels. The conclusion later was that most people watched TV and were not influenced by radio to any great degree. This was a badly judged strategy, which could have been changed to another by some doing some HUMINT.[19] The situation was also not helped by the fact that the Americans, true to practice at home in the USA, broadcast on AM radio, when the Yugoslavs in general—just like in other countries with only FM channels—hardly ever had any cause to switch from their FM radio settings.[20] A border was breached—which resulted in NATO troops confiscating Serb TV transmitters—when clips of SFOR (Stabilisation Force) troops were mixed with old clips of German Nazi troops in the programs.[21] Throughout the whole process, from the start of the war to the midst of it, television was used effectively as a weapon of war by all parties. The purpose of the media campaign was to legitimise the international presence actions in the local Bosnian arena, a strategy that to a great extent focused on the struggle for symbolic capital on the political social field.

Bibliography

Allin, D. H. (2002). *NATO's Balkan interventions*. Oxford: Routledge.

Donia, R., & Fine, J. (1994). *Bosnia and Herzegovina. A tradition betrayed*. London: Colombia University Press.

Gagnon, V. P. (2004). *The myth of the ethnic war. Serbia and Croatia in the 1990s*. Ithaca: Cornell University Press.

---

[19] The Americans who are themselves a people geared to television ought to have realised the importance of the television media. Evidently,the opinion was that the more primitive media of radio was better suited for Yugoslavia.

[20] Wentz (1997, p. 66).

[21] Allin (2002, p. 43).

Honig, J. W., & Both, N. (1996). *Srebrenica: Record of a crime.* New York: Penguin books.

Mann, M. (2005). *The dark side of democracy. Explaining ethnic cleansing.* Cambridge: Cambridge University Press.

Owen, D. (1995). *Balkan Odyssey.* London: Harvest Book.

Pavlakovic, V. (2005). Serbia transformed? Political dynamics in the Milošević Era and After. In S. P. Ramet & V. Pavlakovic (Eds.), *Serbia since 1989. Politics and society under Milošević and After.* Seattle: University of Washington Press.

Sell, L. (2002). *Slobodan Milošević and the destruction of Yugoslavia.* Durham: Duke University Press.

Smith, R. (2005). *The utility of force. The art of war in the modern world.* London: Penguin Books.

Tabeau, E., & Bijak, J. (2005) War related deaths in the 1992–1995 armed conflicts in Bosnia and Herzegovina: A critique of previous estimates and recent results. *European Journal of Population, 21*(2). Heidelberg: Springer.

The Independent International Commission on Kosovo. (2000). *The Kosovo Report. Conflict, International Response, Lessons Learned.* Oxford: Oxford University Press http://reliefweb.int/sites/reliefweb.int/files/resources/6D26FF88119644CFC1256989005CD392-thekosovoreport.pdf [Visited 170509].

Udovicki, J., & Stitkovac, E. (1997). Bosnia and Hercegovina: The second war. In J. Udovicki & J. Ridgeway (Eds.), *Burn this house. The making and unmaking of Yugoslavia.* Durham: Duke University Press.

Wentz, L. (Ed.). (1997). *Lessons from Bosnia. The IFOR experience.* Washington, DC: CCRP Publications.

CHAPTER 10

Conclusions of the Empirical Example

Abstract The discussion above demonstrates that it is possible to acquire useful knowledge for military purposes by studying the prelude to the conflict in Bosnia from a field theory perspective. As seen the method here is more the historian's than one which one would need on a contemporary example. But the case here was to present the use of the theory, not of the methods. As has been said earlier, one can use almost any kind of method when one works with field theory.

Keywords Bosnia · Field theory · PSYOPS

The course of events was described from the strategic down to the operational level as appropriate to the discussion. If the international community had made serious efforts to consider the effects they wanted to achieve, apart from those for a very short term, they would possibly have acted differently. It involves consequence analysis, but also looking at the overall political situation. It was a question of not fully understanding the logic of the field. If one does not understand how the field works then neither will one be able to progress to the next stage and possibly use field theory as a tool. What do the actors want to achieve and how can they be persuaded to change strategy? These questions should serve to direct the handling of this type of situation. An analysis of events

© The Author(s) 2017
H. Gunneriusson, *Bordieuan Field Theory as an Instrument for Military Operational Analysis*, New Security Challenges,
DOI 10.1007/978-3-319-65352-5_10

was conducted using field theory, with the theoretical terms sometimes clearly shown and sometimes indicated through the use of synonyms. Even though the actors did not always think in field theory terms, it is still possible to apply that perspective. To that should be added that much indicates that Milošević and Tudjman, without any actual theoretical knowledge, could think and act in a manner reminiscent of the field theory approach. That is something you can and also probably actually do if you are an actor on the field. The opposite can be said about many of the other actors who really did not try to understand the social field and thus really did not take the duel-situation between them and their counterparts seriously.

One can also see that it was possible to extrapolate the coming events by studying the actors and that their rationality also could be perceived. This is exactly the information one needs if one wants to affect actors by changing the social structure on a field. PSYOPS could be applied right through the whole event chain, not only during the phase when open war was played out. Even a bombing or an assassination of other actors than those you want to affect can be part of a PSYOPS operation. Exactly what could have been done is speculative and beyond the point of this text. The important thing is that it was possible to operate in this setting with a social field perspective as guidance.

The aim has been to show that field theory can fulfil the criteria necessary for a theory to qualify as a valuable instrument for military planning. The discussion has demonstrated that social theory can be employed as an operational tool. The theory can serve to encourage reflection and therefore refine one's thoughts, provided that it is first understood and assimilated as a way of thinking. This provision was fulfilled when the theory was shown to be able to be linked, for example, to the questions and issues that a military intervention might have to face. The theory was also shown to be applicable to the empirical example given. One counter to this claim is of course that it is a *theoretical* examination of empirical data that has been conducted. The next step is to apply the theory in an actual empirical study. But for this to be possible the theory must first be further developed and then practical measures taken to enable theoretical ambitions to be realised—in other words a task outside this study. One question that is inherent and needs to be answered following this type of study is: what resources are required to enable the creation of a field theory image of military area of operations?

As shown above, it might seem an easy thing to identify how the enemy is structured and what his possibilities are on the social field, and by that outmanoeuvre him. But as Carl von Clausewitz stated "Everything is very simple in war, but the simplest thing is difficult".[1] What is the catch then? There are certainly many, but now when we have this perspective laid out in front of us, then we can go one step further. We have already discussed that people are structured to value certain things on a social field and that they are also structured to most likely not do certain things according to their habitus. But these limitations apply to the other side as well, to our forces or our allies. We are structured by our culture, upbringing, regulations, rules and schooling in general to do certain things and not do others. On top of that we are not only becoming an actor on the local political field, we are at the same time acting on a multitude of other social fields. Our forces are, for example, acting on a military social field, with its own rules, values and limitations. These limitations from our own culture leave us severely limited when it comes to fitting our actions towards a local social field, and against the habitus of the local actors (collective or individual ones), or just trying to act together with allies can actually provide a daunting task as collective habitus varies between nations and organisations. Things which might make sense for us might be counterproductive seen from the local perspective. For example, forbidding all Baath party members to be employed by the US administration in Iraq might have been a good idea if it was not almost equal to forbidding anyone having knowledge of administration to have a job within the Iraqi state. Or the initial idea to leave out the high priests of Iraq from the political process was clearly a misunderstanding of the political role Islam had and still has in Iraq.

By way of a start it is suggested that the approach would be interesting to apply to a staff exercise. A number of practical questions need to be examined, which would be best addressed by conducting an exercise in the form of an experiment. There will certainly be some need for the staff to have read up on Pierre Bourdieu, at least one or two of the more solid works he has written, like *Logic of Practice*, The Rules of Art, *The Field of Cultural Production*, or why not his earliest field study on the Kabyle *Outline of a Theory of Practice*. As an example of this, an

---

[1] Carl von Clausewitz (1991). Book 1, Chap. 7.

important type of issue is where the borderline between the operational and the two strategic levels goes when conducting an analysis of the field. In an operational assessment of the field, ne might consider setting priority on a particular process to examine the structure of the field, so that certain actors, who for political reasons should not be engaged, are affected in a negative way. This will of course generate frictions. These frictions, however, lie embedded in the different levels of the structure, rather than as a result of the application of field theory. The latter is more directed towards illuminating problems and provides an overall picture to which the various levels of command can refer. However, exactly how this will happen is an empirical issue best examined through experiment. What the study does bring to light is that field theory will be most effective if the intervention force is in the operations area for a protracted period, when comprehensive information gathering can be conducted and be fruitful in the long run.

BIBLIOGRAPHY

von Clausewitz, C. (1991). *On war.* http://www.gutenberg.org/files/1946/1946-h/1946-h.htm [Visited 170320].

CHAPTER 11

Post Scriptum: Hybrid Warfare and the Yugoslavian Blueprint

Abstract With the recent development of Russian international politics, it is of interest to revisit Yugoslavia and discuss the lessons learnt there in the light of hybrid warfare. The events in Yugoslavia in the 1990s resemble in parts to what has been described as hybrid warfare in Ukraine in 2014–2017.

Keywords Hybrid warfare · Russia · Kosovo · Ukraine

Hybrid warfare has as a term travelled a lot in definition. It started as a description of the Israeli debacle in Lebanon 2006, there it was seen as an empirical theorisation of Hezbollah possessing advanced weapon systems and still not being a state actor.[1] From there, one can speculate what the new hybrid threats can constitute from both future inventions as nanotechnology and biohacking to contemporary cyber threats.[2]

---

[1] Håkan Gunneriusson (2012, p. 49). Much inspiration has been given by Frank G. Hoffman and his work on hybrid threats. Frank G. Hoffman (2009).

[2] Ibid. p. 63 and Håkan Gunneriusson and Rain Ottis (2013).

© The Author(s) 2017
H. Gunneriusson, *Bordieuan Field Theory as an Instrument
for Military Operational Analysis*, New Security Challenges,
DOI 10.1007/978-3-319-65352-5_11

The war in Ukraine which started with the Russian annexation of Crimea can be said to take the term hybrid warfare into another direction.[3]

The Russian occupation and the following annexation of the Crimea peninsula were very much enabled because the large number of ethnic Russians on the peninsula. Some were for the annexation; some were bystanders and some others were against it. But they gave a minimum of legitimacy to the occupation as it could be claimed that the population wanted to secede from Ukraine and join Russia. There is a history to this and it be explained with Yugoslavia as an inspiring scenario. The term hybrid warfare was not conceptualised back then, but it may as well have been forged. There are similarities between the Crimea annexation and some of the elements in the break-up of Yugoslavia. There was an urge to create nation states to replace Yugoslavia which encompassed many different Slavic nations. This can be seen and was also seen as a state building process, more than an example of hybrid warfare. The break-up did not happen by proxy as in Crimea. The cleansing of Knin (*Kninska Krajina*) in Croatia, on the other hand, can be seen through the perspective of hybrid warfare even if the term still was not conceptualised, and the Serbian minority's claim for autonomy from Croatia was unsuccessful as a whole. The Serbs, who undoubtedly had lived in the region for hundreds of years, wanted to secede from the new state of Croatia, they declared themselves autonomous. The support from another new state, Serbia, can be seen as a parallel to the Russian support in Crimea and Russia's proxy involvement there.[4] The difference was that Serbia lacked all resources to effectively include Knin in its war efforts—at least beyond rhetorical statements. Furthermore, there was just an ad hoc plan without further thought of support to Knin from Serbia, in effect not just a lack of capacity but a lack of planning too. The

---

[3] Sascha Bachmann and Håkan Gunneriusson (2017).

Håkan Gunneriusson and Sascha Bachmann (2015a, b). More about hybrid warfare in Håkan Gunneriusson and Sascha Bachmann (2015a, b). Håkan Gunneriusson and Sascha Bachmann (2014). Håkan Gunneriusson and Sascha Bachmann (2017). Amos C. F., and Rossow A.J. (2017).

[4] The lack of land connection between the splinter region and the country encouraging the break up exists in both examples. This is of course not an argument for hybrid warfare, but it can be stated just to remind that the Knin was as much of a satellite of the ethnic mother country as Crimea is.

claim of independence by the local Serbs was based on the fact that the Serbs lived in Knin, there was no doctrine other than nationalistic ethos. Further, Serbia under Milošević lacked all types of symbolic capital to successfully wager such claims for their own part or for that part for Knin as a separate entity. Russia, an active supporter of Serbia then and now, stood by and could nothing do. During Yugoslavia's downfall, Russia was weak both politically and military. All Russia could do was to take notes for their new doctrine while it was a one-off operation for NATO, as with the example of Kosovo discussed below.[5] Did they do that? Most certainly, if not anything else the head of Russia's General staff and the creator of its contemporary doctrine Valerij Gerasimov himself states that Yugoslavia during the 1990s was an example of NATO hybrid warfare.[6]

Another example of a proto-hybrid event is the administrative unit of the Republic of Srpska, which consisted of the eastern part of Bosnia. The Republic of Srpska is by no means recognised as a separate state, it is still a part of Bosnia but has little to do with the rest of the country and the issue is settled as being so for the time being. Autonomy for the area is thus not achieved, nor does Serbia formally control the territory, so this proto-hybrid warfare event was merely semi-successful. If Serbia had the political and foremost comparative military leverage, they could easily have done or yet still do a Crimean hybrid warfare operation there at any time. But being military weak and in the political process of joining the EU there is little to gain by such an approach, they did not have or currently have the capacity to act freely. We have to look elsewhere at the parting of Yugoslavia to find a more hybrid-like scenario. Something classified as a successful hybrid warfare operation can be said to have happened in the example of Kosovo. This is a good example as the process of splitting Kosovo from Serbia was very sensitive for the regime in Russia, and they took it as an unfair stripping of territory from its ally Serbia. This is not dissimilar to the Crimea situation with the USA

---

[5] The mastermind behind Russia's new doctrine is Valery Gerasimov. He presented his ideas in the following article: "The value of science in anticipation. New challenges require rethinking the forms and methods of warfare". http://www.vpknews.ru/articles/14632 [170509].

[6] *VPK News*, number 10 (674) 15th March 2017, abridged version of the report "Modern War and current national defence issues" (authors translation) for the Academy of War studies. http://www.vpk-news.ru/articles/35591 [170509].

backing Ukraine and Russia separating Crimea from Ukraine. In the process of separating Kosovo, it was a stroke of symbolic violence that NATO delivered to Russia, displaying Russian weakness in the Balkans and overall. It also presented Russia with the pretext for its operation in Crimea, the sword of seceding showed to be two-edged and thus useful for Russia as well. Kosovo was staged carefully by the USA and her allies and backed both by civilian tools represented by the UN as well as the OSCE, including the military tool—NATO. Pushing for a greater self-governing of Kosovo initially, NATO presenting a peace treaty to the Serbs. However, this treaty included a paragraph that certainly would be hard for any state to accept as it included free movement and bases by a foreign military alliance (NATO) on the sovereign soil of Serbia and free movement through all of Serbia for NATO-troops. This paragraph ensured that the treaty failed and opened for military aggression. It can be argued back and forth if this paragraph really mattered, if it was raised by the Serbs or not in the negotiations.[7] What is not debatable is that the paragraph was in the agreement refused by the Serbs. Refusal by the Serbs led to aggression on part of the international community with a few expectations and after a 10-week long ordeal Serbia finally agreed to the terms of NATO. The United Nations and OSCE quickly deployed and started the building of local and central institutions for governing of a state.[8] Furthermore, the legal framework was more or less completely changed in the nine years that followed until the Kosovo Albanians declared unilateral independence, backed on the way there and after by NATO as led by the USA.[9] The *illusio*, the rules of the field of international law, was changed with Kosovo, and the change was then confirmed with Crimea when two major international actors on the field, NATO and Russia, had used this new *logic of practice* for international law.[10] Important to note is that the actions taken to separate Kosovo from Serbia was an ad hoc solution to a problem limited in time and space. Nothing else came out of this new way to interpret international

---

[7] http://news.bbc.co.uk/2/hi/europe/682877.stm [170509].

[8] http://www.osce.org/ [170509].

[9] The change of the legal system and various institutions were facilitated by the United Nations and the OSCE through the backing by member states, as well as aid provided by the EU, USAiD and DiFiD to just name a few.

[10] Instead of *new* one can see it as a reversal of the old practice of Realpolitik.

law. Russia took notes and saw a new pattern in this, a new way to act and added it to its foundation for contemporary warfare. They re-coded their understanding of warfare so that they could have their way with their neighbours like Georgia and Ukraine, primarily while still claiming deniability. One can with right argue that Russia also lacked symbolic capital in their claim to Crimea, as for Serbia in the Knin example above. But Russia could point to a host of factors in favour of Russia. Among them the claim that Crimea historically (pre-Stalin) was Russian, along with the ethnic presence of Russians there. More importantly, the ability to pursue a narrative like the one of Putin's Russia can be fruitful from their perspective. The loss of Kosovo for the Serbs can be seen as both as a material loss and a symbolic loss. Russia used the narrative from the case of Kosovo, even if not openly stating it, as a legitimating tool for the annexation of Crimea. It helped them in their effort to build the required minimum of legitimization for their annexation, at least they can hold an argument for it. Furthermore, Russia's massive military supremacy in the region talks for Russia's success as there is not really much to do in military terms about Russia's activities in Ukraine; realpolitik beats finesse in many cases, as in this example. That is the reason why the EU and NATO do not stand up for conventions against war of aggression in the case of Ukraine. The West do not have the capability to challenge Russia in the region, so it is easier to look the other way. It can be seen as a case of *Reflexive control*. This is an old Soviet approach called reflexive control. The psychologist Vladimir Lefebvre defines *reflexive control* as "a process by which one enemy transmits the reasons or bases for making decisions to another".[11] The concept of reflexive control can in fact be used with bourdeiuan field theory as approach as there are similarities shown in this text. The author Timothy Thomas means that reflexive control can be used on all levels of warfare.[12] One of the most complex influence operations is to influence a state's decision-making process.[13] Russia's warfare against the West can be described as reflexive control, *resulting* in hybrid warfare.

---

[11] Quoted in Timothy (2004. p. 238).

17, 2004. https://www.rit.edu/~w-cmmc/literature/Thomas_2004.pdf.

[12] Timothy (2004. p. 239).

17, 2004. https://www.rit.edu/~w-cmmc/literature/Thomas_2004.pdf.

[13] Ibid, Thomas, L. Timothy.

In the case with EU and NATO vs Russia today, we have a difference compared to the cold war; even if we do not believe the Russian narrative at all, we still do not challenge it fully and call Russia out for waging a war of aggression in Ukraine. Nevertheless, Russia's warfare against Western power can be described as reflexive control, *resulting* in hybrid warfare in the meeting with the Western powers. It all depends on who is listening, but Russia clearly sent a message to all in the region and beyond, no matter what they thought about it. The Russian doctrinal stance of hybrid warfare is now operational, and we have yet to see the end of it.

REFERENCES

Amos, C. F., & Rossow, A. J. (2017). *Making sense of Russian hybrid warfare: As of the Russo-Ukrainian war*. Arlington: The Institute of Land Warfare.

Sascha Bachmann & Håkan Gunneriusson (2014). Terrorism and cyber attacks as hybrid threats: defining a comprehensive approach for countering 21st Century Threats to Global Peace and Security. *Journal for Terrorism and Security Analysis*, Syracuse University, 26–37.

Sascha Bachmann & Håkan Gunneriusson. (2015a). New Threats to Global Peace and Society. In *Scientia Militaria - South African Journal of Military Studies*. 77–98.

Sascha Bachmann & Håkan Gunneriusson. (2015b). Russia's Hybrid Warfare in the East: Using the Information Sphere as Integral to Hybrid Warfare. In *Georgetown Journal of International Affairs - International Engagement on Cyber V: Securing Critical Infrastructure*. 199–211.

Gunneriusson, H. (2012). Nothing is taken seriously until it gets serious: Countering hybrid threats. *Defence Against Terrorism Review*, 4(1). 97–108.

Gunneriusson. H., & Ottis, R. (2013). Cyberspace from the hybrid threat perspective. *The Journal of Information Warfare*, 12(3). 97–108.

Gunneriusson, H., & Bachmann, S. (2017 upcoming). Western Denial and Russian control. How Russia's national security strategy threatens a Western-based approach to global security, the rule of law and globalization. *Polish Political Science Yearbook 2017*.

Hoffman & Frank. (2009). Hybrid warfare and challenges. *Joint Forces Quarterly* (52). 34–40.

Timothy, T. L. (2004). Russia's reflexive control theory and the military *Journal of Slavic Military Studies, 17*, 238. https://www.rit.edu/~w-cmmc/literature/Thomas_2004.pdf [170509].

BIBLIOGRAPHY

Sources

Gace, N. (1991 November 27). Velika Srbija na mala vrata. *Vreme.*

Isakovic, Z. (1991 December 16). Rovosi u dusi. *Vreme.*

Mesic's, S. (2001). Witness at Hague tribunal, *Tribunal update* 68. 16–21 March 1998 in Naimark, N.M. *Fires of Hatred. Ethnic Cleansing in Twentieth-Century Europe.* Presidents and Fellows of Harvard College: Boston.

Nincic, R, et al. (1991 October 21). "Drina bez Cuprije", *Vreme*, Belgrad.

Osland, K. M. (2005). The Trial of Milošević. In S. P. Ramet, & V. Pavlakovic (Eds.), *Serbia Since 1989. Politics and Society under Milošević and After.* University of Washington Press: Seattle.

Szporluk, R. (1988). *Communism & Nationalism. Karl Marx versus Friedrich List.* Oxford: Oxford University Press.

VPK News #10 (674) (2017 March 15). http://www.vpknews.ru/articles/14632 [Visited 170509]

Bibliography

Bourdieu, P. (1977). *Outline of a theory of practice.* Cambridge: Cambridge University Press.

Bourdieu, P. (1984). *Distinction. A social critique of the judgement of taste.* Boston: Harvard University Press.

Bourdieu, P. (1993). *The field of cultural production. essays on art and literature.* Cambridge: Cambridge University Press.

© The Editor(s) (if applicable) and The Author(s) 2017
H. Gunneriusson, *Bordieuan Field Theory as an Instrument for Military Operational Analysis*, New Security Challenges, DOI 10.1007/978-3-319-65352-5

Biljelic, B. (2005). Nationalism, motherhood, and the reordering of women's power. In S. P. Ramet & V. Pavlakovic (Eds.), *Serbia Since 1989. Politics and Society under Milošević and After.* Seattle: University of Washington Press.

Broady, D. (1998). http://www.skeptron.uu.se/broady/sec/ske-15. [Visited 170320]

Callan, M. & Ryan, M. (2005). Effects- based operations: Discussion paper. In J. Elg. (Ed.), *Effektbaserade operationer.* Stockholm: Swedish National Defence University.

Coram, R. (2002). *Boyd: The fighter pilot who changed the art of war.* New York: Back Bay Books.

Dahl. E. J. (2002). Network Centric Warfare and the Death of Operational Art. In *Defence studies, 2*(1). Oxford & New York: Routledge.

Deptula, D. (2001). *Effects-based operations.* Aerospace Education Foundation: Arlington.

Headquarters Department of the Army. (2008). *FM 3-05.130 Army Special Operations Forces Unconventional Warfare.* Headquarters Department of the Army: Washington, DC.

Heinecken, L. (2015). The military and Society: The need for critical sociological engagement. In *Scientia Militaria, South African Journal of Military Studies, 43* (1) (Stellenbosch).

Kitson, F. (1989). *Directing Operations.* London: Faber and Faber.

Malešević, S. (2010). How pacifist were the founding fathers? War and violence and classical sociology. *European Journal of Social Theory, 13*(2). London: Sage.

Sun, Zu. (1997). *Sun Zis krigskonst.* Santérus: Stockholm.

Virilio, P. (2005). *Desert screen. War at the speed of light.* London: Continuum.

Walby, S. (2012). Violence and society: Introduction to an emerging field in sociology. *Current Sociology, 61*(2).

INDEX

A
Actor, 4–6, 11, 14–21, 23, 24, 28–32, 36–40, 42–44, 47, 51, 53, 54, 57, 58, 60, 66, 70, 71, 78, 79, 83, 90, 91, 96, 107–109, 111, 114
Arkan, 46, 81

B
Baker, James, 77
Bildt, Carl, 94, 103
Bosnia, 30, 46, 51, 56, 64, 71, 76, 78, 80, 81, 84, 85, 89–91, 93, 94, 96, 100–103, 113
Both, Norbert, 101
Bourdieu, Pierre, 3, 10, 18, 109
Browning, Christopher, 46, 60

C
Callan, Michael, 8
Capital
    cultural, 17
    social, 23, 31

symbolic, 22, 23, 28, 31, 32, 55, 66, 67, 92, 102, 104, 113, 115
Chetnik, 56, 61, 67, 70, 80, 81
Clausewitz von, Carl, 109
Corum, S. James, 22
Creveld, van, Martin, 21
Croatia, 55, 57, 59, 63–71, 76, 78, 80–83, 85, 89, 90, 93, 95, 100, 101, 112

D
Drašković, Vuk, 58, 61, 70, 81
Dulić, Tomislav, 67

E
EBO, 3, 7, 8
Effects-Based Operations. *See* EBO
EU, 77, 78, 81, 85, 89, 91, 96, 113, 115, 116

F
Faurisson, Robert, 22

© The Editor(s) (if applicable) and The Author(s) 2017
H. Gunneriusson, *Bordieuan Field Theory as an Instrument for Military Operational Analysis*, New Security Challenges, DOI 10.1007/978-3-319-65352-5